WHEELS OF FARM PROGRESS

Marvin McKinley

 American Society of Agricultural Engineers

FIRST EDITION

LCCN: 80-68925
ISBN: 0-916150-24-0

Printed in the United States of America by the American Society of Agricultural Engineers, 2950 Niles Road, P.O. Box 410, St. Joseph, Michigan 49085.

Table of Contents

Foreword . . . . *iv*

About the Author . . . . *vi*

It All Began With the Reaper 1

The Glorious Days of Steam Threshing 25

Those Hit and Miss "Hired Hands" 53

Harnessing the "Modern Farm Horse" 69

From Leather Reins to Steering Wheels 97

Getting the Farmer Out of the Mud 111

The "Devil Wagon" Invades the Farm 127

Bibliography 154

These two early-day photos were taken about 1906 or 1907. The engineer standing by the smoke-belching Aultman & Taylor in the threshing scene is the author's great-uncle, Lewis McKinley. His brother, Willard (in the dark shirt), feeds the separator. The identities of the rest of the crew are unknown.

Foreword

Life on the farm during the 1930's was an unforgettable adventure for me. As a youngster in grade school, every activity throughout the year had a special meaning. There was spring plowing and planting, haymaking, grain harvesting, fall plowing and planting, and finally, corn harvesting. Although tractor power was used for some of the heavy work, most activities were performed by horses. My father had four of them on our farm.

Threshing was unquestionably the biggest and most eagerly-awaited event of the season. The sight of that formidable rig rounding the bend of the road, lumbering slowly toward the lane that entered our farm, was a thrilling prelude to the activities which were about to begin. I always watched spellbound the spectacle of bundles moving endlessly on the self-feeder, the grain streaming down the bagging spout, and the straw shooting out of the blower and settling onto the stack.

Providing the power for this operation was an aging "OilPull" tractor. To me, this behemoth was awe-inspiring, with its mighty two-cylinder engine, ponderous iron flywheel and towering exhaust stack. I recall it was sometimes difficult to start the tractor by the usual cranking method, and that on such occasions the entire crew grabbed hold of the long drive belt, straining in unison to turn over the engine.

Attending this rig was a custom man somewhat small in stature but large in threshing know-how. Among my memories of him, the one which perhaps stands out most vividly, was his fondness of pumpkin pie. My mother always made certain that at least one such pie was served at the threshing table.

There was only one disappointing thing about threshing time, and that was its duration. Before I knew it, the grain bins were filled and the stack was "topped out." The event was over for another year. I always watched regretfully as the outfit moved out of the farmyard and down the lane, bound for the next job. I was certain that some day I would become a thresherman.

By the time I reached manhood, however, the farming methods which I had come to know and admire were fast disappearing. The combine was ringing the knell of the time-honored threshing machine. Old Dobbin, whenever still found, was making his last stand. Customs which had been practiced for generations were being cast aside.

Unmistakably, it was the end of an era. But what a significant, dramatic era! I was saddened to realize that with its passing, life on the farm would never be the same again.

It was the desire to record some of the history and romance of this bygone era for future generations — and to recall fond memories of it for others — that prompted me to write this book.

May it warm alike the hearts of the young and the not so young.

Marvin McKinley

Holding the team at left in the farmyard scene is the author's grandfather, John Obrecht. Kneeling is his son, Glenn. Standing at center is the author's great-grandmother, Caroline Garst (born in 1845). Holding the team at the right is her son, John. The horses are Queen, Doll, Tom and Jim.

About the Author

Born and reared in a farming community in north-central Ohio, Marvin McKinley, as a boy, was fascinated by the prevailing style of rural living. Tales of earlier times, related by his grandparents and others, instilled in him a keen interest in agricultural history. The sweeping changes in farming methods which took place during the 1940's brought an era in American agriculture to a close. McKinley considered this to be in many respects regrettable, and conceived the idea of someday perpetuating the era by recording its colorful events.

During the past 15 years he has devoted a great many hours to the study of books on early-day agriculture, rare farm machinery literature and other related material. In 1976 he completed the major portion of his research and began the writing of this book. *Wheels of Farm Progress* is the result.

Marvin McKinley

Other Books in ASAE's Historic Series

Advance-Rumely OilPull Instruction Manual
The Agricultural Tractor: 1855-1950
Farm Power (reprint of 1915 IHC book)
The Grain Harvesters
Rural America A Century Ago
Travel Historic Rural America

1
It All Began
With the
Reaper

In his hour of triumph, Cyrus H. McCormick strides behind his revolutionary reaper during its successful trial in July, 1831. A servant raked the cut grain from the platform (International Harvester Co.)

Living in a vast domain, the early American farmer had all the land he could work, and more. When man power could not supply his needs adequately he relied upon ingenuity, and by utilizing animal power he multiplied his own efforts.

Over the years, many mechanical aids were developed which steadily lightened the work load of the farmer. By 1870 more than 2,000 firms in the United States were manufacturing some sort of agricultural machinery.

An expanding population, with its ever-increasing demands for food, became the farmer's relentless taskmaster, requiring more land and more effective means of working it. But eventually the evolution of agricultural machinery slowed down, due to limitations inherent in draft animals. By the end of the 19th century, many implements had been perfected to the point where only additional power could increase their efficiency.

Harvesting grain with a cradle was considered an art. In some communities cradling was almost a trade by itself. A good cradler received two or three times as much pay as an ordinary farm hand. (Ohio Historical Society)

Although the steam engine left an indelible mark in the wake of rural progress — most notably for its contributions to grain threshing — it was the internal-combustion engine that provided the broad application of power needed to advance farming methods. As the gasoline tractor gradually displaced the horse, farm machinery was redesigned for greater speed and capacity to complement the new form of traction. Agriculture and power machinery had become inseparable.

It all began on a midsummer day in 1831 near Steele's Tavern, Virginia. The stillness of the countryside was about to be broken by a public demonstration that would mark the beginning of a new epoch in agricultural invention. The small crowd of bystanders was curious and skeptical, but as Cyrus H. McCormick's new creation moved down the field the wheat fell in a steady stream upon its platform. The whirling gears of the mechanical reaper soon would be a familiar sound in the American harvest field.

During the harvest of 1833, an Ohioan named Obed Hussey operated a successful reaper of his own design. His invention was unique in that it could be converted from a reaper to a mower by simply removing or adjusting certain parts.

Until the advent of the reaper, advances in farming methods had been so slow they were hardly perceptible. Oxen provided much of the power for heavy work, but most operations were done by hand. The cradle was used universally for harvesting grain. Essentially, it was a scythe with several wooden fingers attached parallel to the blade. These fingers placed the stalks in a swath for easier binding. With a cradle, a strong man could cut two or three acres of good standing grain in a day. Another man, equally able, was required to rake and bind it.

Although the early reaper could cut eight or 10 acres in a day, it offered no relief in raking and binding. It required a man with rake in hand to keep the platform clear of cut grain, and several men to bind it into sheaves.

Improvements came in the 1850's, first with a self-rake device which automatically swept the grain from the platform. This released the man with the hand rake for other work. In 1858, the Marsh Brothers of De Kalb, Illinois, introduced a harvester which employed continuous canvas aprons. The grain was elevated over the drive wheel and delivered to two men who, while riding, performed the binding chores.

In 1871, a self-binding device that tied the bundles with wire was developed. Although manufactured in considerable numbers, the day of the wire binder's supremacy was brief. Farmers complained that the wire became scattered all over their farms and often was swallowed by

The self-rake reaper swept the cut grain mechanically off the platform and deposited it in neat piles ready to be bound. A crew of four or five men was required to do the binding. (International Harvester Co.)

livestock. Four years later, John F. Appleby invented the first successful twine knotter. It underwent various improvements until it was perfected in 1879. This device became the foundation of the binding mechanism used on nearly every binder built thereafter.

In the annals of labor-saving machinery, perhaps no implement rendered a greater contribution to the march of agricultural progress than the self-tying grain binder. This machine reigned supreme in the American harvest field for 60 years. Its initial impact upon agriculture transcended that of any implement introduced in the 19th century. Production of harvesting machines rose from 60,000 in 1880 to 250,000 in 1885.

When the self-binder was first put on the market, farmers considered it a modern wonder. However, in the hands of unskilled operators, the problems which arose seemed endless. The binder was the first really complicated machine to be used by the farmer, and its successful operation demanded a greater knowledge of mechanics than he possessed. The assistance of many trouble-shooters was needed each season in cutting and binding the nation's grain crop.

The advent of the self-binder marked the beginning of a new era in grain harvesting. With machines such as this Champion, one man could perform the entire cutting and binding operation.

But during the ensuing years, binder construction improved steadily. Its universal adoption made it a familiar object on nearly every farm and a subject of discussion wherever farmers met. A new generation grew up amidst the binder, one that learned its principles in the years of early youth when the mind was most receptive. Eventually, mechanical difficulties became almost a thing of the past.

All binders were similar in general construction and principles of operation. The combined action of a reciprocating sickle-bar and rotating reel severed the stalks and gently pushed them backwards onto a moving canvas that covered the binder platform. The stalks then were conveyed upward between two elevator canvases and delivered to the binding attachment. This mechanism, consisting chiefly of a needle and knotter, bound the grain into convenient-sized sheaves, then discharged them onto a bundle carrier. The carrier, holding from four to six bundles, was tripped at regular intervals around the field, thus forming rows to make easier the job of shocking.

Above: The Marsh Harvester possessed the same cutting mechanism as the early reapers, with an elevator and binding platform added. Two men rode on the platform and bound the grain by hand. (International Harvester Co.)

Below: The McCormick binder was a favorite in every grain-growing community. Its design, as well as the design of all other binders, remained basically unchanged over the years. Only minor improvements, tending to give greater durability and lighter draft, were occasionally added.

Grain binders were manufactured in sizes to cut six, seven, or eight-foot swaths. An eight-foot machine, drawn by four horses, could cut from 15 to 18 acres a day, allowing time for lubrication and resting the animals.

This Deering grain binder of 1920 had an eight-foot cut. Like all machines of this size, it required the use of a four-horse team. Deering binders were especially noted for their light draft.

A large field of shocked oats in the 1930's awaits the trip to the threshing machine.

It was common practice to oil the entire binder before starting out in the morning and every three or four hours thereafter. To operate an eight-foot machine continuously, day-in and day-out, required four additional horses. Teams were changed at noon, and if required in mid-morning and mid-afternoon, as well.

The proper time to cut wheat with a binder was when the straw just below the head and above the last leaf was a golden yellow. If the wheat was harvested before this stage, there often were many shriveled grains. On the other hand, if the grain was allowed to fully ripen before being cut, many kernels were shattered and lost.

Grain had to be well-shocked soon after it was cut and bound. Serious losses could result if this work was postponed or done carelessly. The ideal shock was one that would stand up in windy weather, ensure quick and uniform curing, and afford the grain heads as much protection as possible from the weather. Eight or more bundles were set with butt ends on the ground, tops leaning in and meshed together. To shed the rain, a cap bundle was placed on top with its heads spread fan-wise in the windward direction.

Harvesting grain with a cradle became a rare sight soon after the introduction of the self-binder. However, some farmers continued to use the cradle for many years to "open up" their fields. This practice prevented losses from trampled grain that otherwise would occur when

When the tractor hitch was used, several binders could be operated at once, thus saving much valuable time on big-acreage farms. The tractors in this scene are OilPulls.

the team had to walk in standing grain during the first round. Hand methods also were used occasionally in times of emergency. In 1902, a local newspaper in an Ohio Community reported: "The heavy rains during the past week caused many of our farmers to turn back towards the old primeval days and go into their fields with cradles. Some bought new 'out of dates.'"

In 1909, a Minnesota firm introduced a simple but ingenious attachment—a tractor binder hitch. With this hitch, a tractor could draw as many binders as its power capacity would allow. The hitch was arranged so that the attendant of each binder could control his machine independently of the others. By pulling a rope secured to a lever on the tractor and threaded through the line holders of each machine, he could stop the tractor instantly. Independent steering permitted adjustments and minor repairs to be made while moving. In such instances, the attendant of a disabled binder steered it over behind the one to his front, made his adjustment, then steered the machine back into the grain to take up his swath. While the disabled binder was out of operation, the machines that followed it were moved over a swath so that no grain was missed.

With four eight-foot binders about 65 or 70 acres could be cut in eight hours. A field of standing grain quickly became a spectacle of stubble and eye-appealing shock rows.

Grain binders designed especially for tractor use appeared in 1919. These machines were made in eight and 10-foot sizes and operated at faster speeds than horse-drawn models. A 10-foot tractor binder could harvest almost as much grain in a day as two eight-foot horse binders. Power was transmitted to the machine by means

Opposite: This novel method of cutting grain was offered by Deere & Company in the early 30's. A ten-foot tractor binder with "steering control" gave one man complete command of the operation from the tractor seat. (Deere & Co.)

of a power take-off shaft, which eliminated the need for a main drive wheel. This made it possible, on soft ground or in heavy grain, to maintain the speed of the cutting and binding mechanism while reducing the speed of travel.

More than 150,000 grain binders were being manufactured annually when the one-man combine, then called the combined harvester, appeared in the mid 1920's. Competition from this new machine increased steadily, but did not become formidable for another decade. An estimated 1,250,000 binders still were in use on American farms in 1938.

By 1837 many farmers in Illinois were giving up in despair. The cast iron plows they had brought with them from the East were satisfactory for breaking the virgin sod, but after the first plowing the heavy prairie soil stuck to the moldboards and had to be scraped off every few steps. Discovering that scraping required as much time as actual plowing, many considered abandoning their land and moving farther west.

In the town of Grand Detour, Illinois, John Deere, a young blacksmith, was studying the problem. He selected a one-piece wrought iron plow and, using a discarded mill saw blade, carefully welded a cutting edge of steel to the moldboard. The success of this "self polisher" was phenomenal. In addition to its scouring qualities, the steel plow's lighter draft made turning the soil at higher speeds possible.

Eighteen years after Deere designed his first plow, another blacksmith, James Oliver, purchased a quarter-interest in a foundry in South Bend, Indiana. He was convinced he could develop a better plow that could sell for less money, so he set out to improve the cast iron plow. Two years later he patented a method of "chilling" plowshares and in 1868 succeeded in chilling the moldboard, producing a hard surface that resisted wear and a soft core that resisted shock.

The ever-present walking plow was manufactured in a variety of forms. A farmer had his choice of either wooden or steel beams and either left- or right-hand moldboards. The 12-inch share was generally considered standard for use with two horses. However, sizes ranged from seven to 16 inches, adapted for from one to three horses.

While both steel and chilled bottoms were used extensively, some farmers preferred "combination" bottoms. The latter combined the lightness, strength and wearing qualities of a steel moldboard with the cheap and easily renewable features of a chilled share.

Plow bottoms were available in many different styles to meet the demands of different soil conditions. In 1916, Deere & Company offered a prospective buyer nearly 1,000 options. But for all practical purposes, plow bottoms could be classified under three general heads: old ground or stubble bottoms, general-purpose bottoms, and breaker bottoms.

7

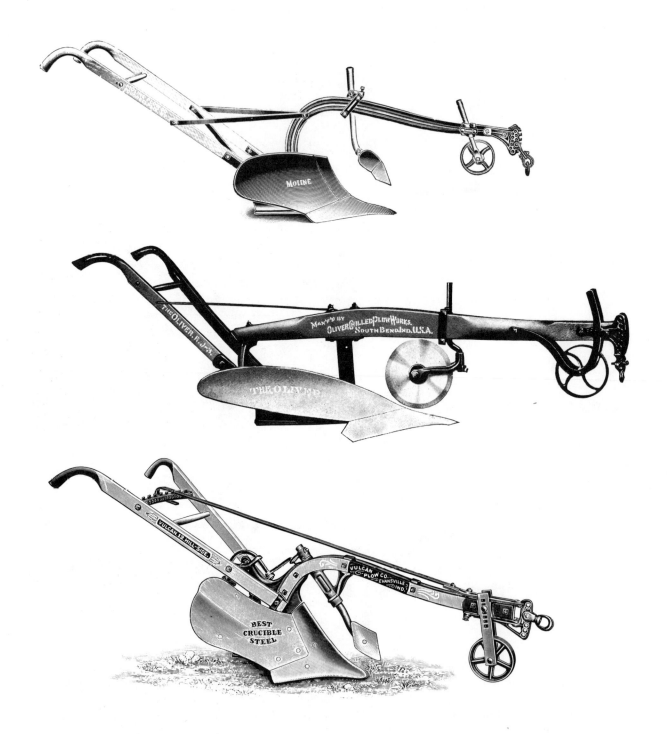

Stubble bottoms featured short, high moldboards with an abrupt turn for pulverizing the soil. They were principally designed for farmers who grew small grain crops. Having a slower moldboard turn, general-purpose bottoms were adapted especially for farmers with a crop rotation that included sod as well as stubble. Breaker bottoms had a long, tapering moldboard with a slow, easy turn which made them light in draft. As the name implies, they were used in breaking unsubdued lands, usually virgin sod or recently-cleared ground. Several additional variations were designed to meet a wide range of requirements. Sandy land, black land, mixed land, middle breaker, brush breaker, prairie breaker, new ground, and hillside bottoms all were popular.

Above: General-purpose walking plows were designed to turn a neat, well-crowned furrow under a variety of soil conditions. This Moline was said to do good work whether in stubble, sod, clay or sandy land.

Center: Breaking plows were built in several different styles, each being designed for a particular type of work. This Oliver was used to break virgin sod. Note the unusually long moldboard.

Below: Hillside plows could be reversed without stopping the team. A slight touch on a foot pedal released a lock latch, which allowed the base to swing from one side to the other. This example is a Vulcan.

Above: Plowing was always a tiring job for a team of horses. The animals in this 1920 scene are pulling a Chattanooga, one of numerous walking plows of the period.

Below: Straight furrows were turned in this circa World War I scene. Walking plows with 14- and 16-inch bottoms required the use of three horses.

Hillside bottoms were unique in that they swiveled from one side to the other, permitting either left or right-hand plowing. By reversing the bottom at each end of the field, the user could plow back and forth on the same side of the land, thus throwing the furrows all in the same direction — always downhill. Plowing in this manner not only reduced draft but also eliminated dead furrows — a major cause of erosion on sloping ground. In many hilly sections of the New England and Middle Atlantic states, this was the standard method of plowing.

The Gilpin two-wheel sulky, introduced by Deere & Company in 1874, could plow up to three acres in 12 hours. It was the first successful plow on which the farmer could ride. (Deere & Co.)

The three-wheel sulky, developed in the middle 1880's, possessed many desirable features not found in the earlier two-wheel models. This Oliver No. 11 was considered the most popular sulky plow ever built.

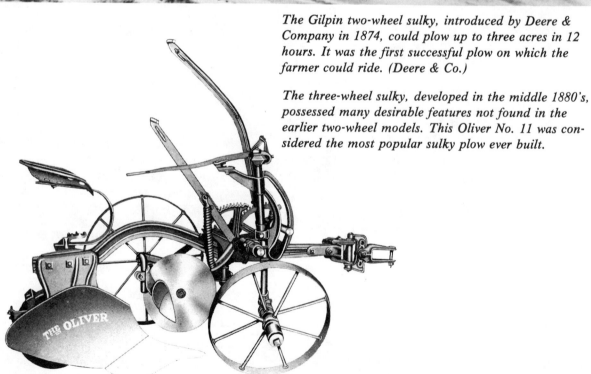

Gang plows appeared in considerable numbers around the turn of the century. These implements nearly doubled the farmer's work capacity compared with the sulky. This gang plow is a New Deere. (Deere & Co.)

In 1874, Deere & Company introduced the first successful plow on which the farmer could ride. It was the two-wheel sulky plow. When the three-wheel sulky appeared about 10 years later, riding plows were commonplace. By use of an inverting land-wheel axle, this implement could remain level while in or out of the ground. On several models the rear caster wheel acted as a rolling landside, reducing the draft substantially.

Another characteristic peculiar to the sulky plow was its ability to turn square corners without being removed from the ground. The point of the plowshare was a pivotal point, around which the land and furrow wheels traveled in a circle. This feature enabled the farmer to plow around his fields — from corner to corner — thus eliminating the time spent in crossing each end with the bottom raised, as was necessary when plowing in lands. By turning either "gee" or "haw," furrows were thrown in one year and out the next.

The gang plow was basically the same construction as the sulky, except that it had two bottoms. With a four-horse team, it could turn over four or five acres in a day.

In 1899 manufacturers added a foot lift, which raised or lowered riding plows with a minimum of effort. When in motion the team, assisted by a strong lifting spring, supplied the power.

Plow manufacturers frequently used this new device as a selling point in their advertising. To emphasize the lifting power of their plows, the Emerson-Brantingham Implement Company illustrated a 110-pound man raising a 14-inch gang with the added weight of a 506-pound man standing atop the implement.

In the early 1900's the horse plodded along undisturbed ahead of the plow. Turning of the soil of more than 200 million acres was an annual task that summoned the utmost resources of the American farmer and his draft animals. Working day-in and day-out, there was only one pace that could be maintained — about two miles per hour. But because of turns and stops for rest, a mile and a half of net work per hour was all that could be expected of a plow horse. Following Old Dobbin behind the walking plow often seemed like an interminable job. A plowman covered more than eight miles while turning over each acre with the average-size bottom. The services of two teams and two men were needed each season to do the plowing and other work of a 100-acre farm.

This scene of all-purpose tractor and two-bottom plow was taken about 1916. The transition from animal to *tractor plowing on the average-size farm was getting under way at this time.*

The year 1915 saw manufacturers turning their attention to the development of implements for use with the new all-purpose tractor. The introduction of two and three-bottom plows, controlled from the tractor seat, added a new dimension to power farming and spelled the decline in use of the plow horse. The transition from animal to tractor plowing was steady but not rapid. For generations field work had been done with horses, and it was difficult for some men to think power farming even after the long and successful use of tractors.

BLACK HAWK
SPREADERS

Hauling manure with the ordinary farm wagon was becoming old fashioned in 1915. One man with a manure spreader could do the work of several men spreading by hand. This implement is an Oliver "Blackhawk."

Look at this Picture! Can a man double disk his ground, and leave it as you see it here, without getting a bigger, better yield? It will pay you to do likewise. A new Keystone tandem disk harrow is waiting for you at the dealer's.

Disk harrows gave deep penetration in hard clay soils. They were manufactured in both single and tandem styles, in sizes requiring from two to five horses. This tandem harrow is a Keystone.

Once the ground was plowed, it had to be "fit." The spring tooth harrow appeared in 1869. It rapidly supplanted the triangular or "A" shaped harrow then in use. With teeth of spring steel, it represented a vast improvement in tillage methods, virtually eliminating frequent tooth replacement in stony soils and in "new ground" with its annoying roots.

Although appearing almost simultaneously with the spring tooth, the disk harrow was not initially as popular as its counterpart. But by the 1890's its use rose appreciably, particularly in the heavy clay soils of the corn belt. Its ability to cut and turn under stubble and trash provided a degree of versatility not inherent in other types of tillage implements.

Above: Spring tooth harrows did excellent work in a variety of soil conditions. They were especially desirable during cold spring weather because their penetrating action left the soil in good condition for absorbing the sun's heat.

Below: The Hoosier grain drill in this 1913 North Dakota scene was exceptionally large for its day. Implements with 10 to 12 furrow openers were standard on average-size farms.

The grain drill was coming into general use in the 1870's, but it was still common to see the farmer with knapsack over his shoulder, sowing his grain broadcast. Some sowers used one hand and covered a strip about eight feet wide. The more expert, however, used the double broadcast, with seed bag hung directly in front and broadcasting with both hands.

16

The chief requisite in a check-row corn planter was accuracy. Any more or less than three kernels in each hill usually meant a reduction in yield. The corn planter in this 1930's scene is a John Deere No. 999.

Many farmers found a combined riding and walking cultivator desirable. The example shown was manufactured by the Moline Plow Company.

With hand methods it was possible to sow up to 20 acres a day, but it was difficult to get an even stand. The problem was in controlling the depth; although sown uniformly, part of the grain was never covered. Sowing with a grain drill gave a better stand, used about a third less seed per acre, and eliminated the need to harrow the field after planting. Tests revealed that using a drill increased wheat yields by three or four bushels per acre, and oats by up to nine bushels per acre.

Marking out rows and digging hills by hand had always been a tedious job. In 1860 George W. Brown was issued patents on the first successful two-row corn planter. To operate it required the services of a man and a boy. The man drove the team and the boy tripped a lever for cross-checking. During the 1870's an automatic check-rower was introduced. This device dropped the seed in hills, thereby permitting cross-cultivation — the favored method of eliminating weeds throughout the horse-drawn era. Checking originally made use of a long, knotted cord, anchored at each end of the field, which tripped the planter's mechanism at the proper time. In 1880 this cord was replaced with wire.

"Corn plowing" always came at haymaking time — one of the busiest times of the season for the farmer. Early methods of cultivating corn were both time-consuming and laborious. The manpower shortage created by the Civil War stimulated the development and use of many labor-saving devices, one of which was the sulky or riding cultivator. Many early models had both a seat for riding and handles for walking. The sulky cultivator doubled the amount of corn that could be conveniently tended — usually from four to six acres a day.

Before the introduction of hybrids, the selection of seed corn each autumn was a necessary chore on every farm. With open-pollinated varieties such as Reid's Yellow Dent and Boone County White, a farmer would roam his fields with a sack and select the desired ears from the best stalks. This was done before cutting time, just after the husks had turned yellow. When seed was selected as the corn was harvested, as was sometimes the case, a padded box was attached to the side of the wagon to contain those ears chosen. Then the ears were cured by hanging them in a well-ventilated, rodent-proof building.

Horse-drawn mowing machines changed little in appearance over the years. This mower was manufactured by Emerson Talcott & Company in 1888. Although it was popular with this 4½-foot cutter bar, larger sizes were available.

Opposite: When using a self-dump rake, all the operator had to do was press a small foot lever and the horses did the rest. Once raked, the windrows were loaded onto wagons with pitchforks. This Osborne rake had a patented whiffletree which aided the horse in turning.

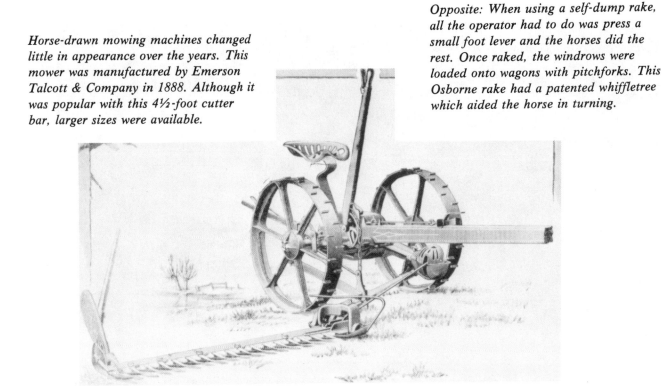

Deering mowing machines of the early 20's were built in three types: regular lift, vertical lift, and giant. The regular lift shown here was the type commonly used on most farms. Giant mowers were so named because they cut wide swaths and had a heavier frame and wheels.

18

Most side delivery rakes of the horse-drawn era actually were two machines in one. By simply shifting a hand lever, they were converted quickly into a tedder. The rake illustrated is a John Deere.

Meanwhile, the development of new machines for use in haymaking was taking place. The elimination of hand methods in cutting the hay was nearly complete by 1865. A mowing machine had appeared which differed little from later models, aside from a wooden frame. Sizes normally ranged from four and one-half to six feet, but a seven-foot machine eventually became available. It could cut up to 20 acres a day. Several manufacturers offered a choice of gear or chain-driven models.

Earliest mowing machines had a hand lever for lifting the cutter bar over obstructions such as stones and stumps, but about 1890 a more convenient foot lift appeared. For farmers with unusually rough or partially-cleared fields, a "vertical lift" mower was manufactured. The cutter bar on this machine could be raised from a horizontal to a vertical position without the operator leaving the seat or stopping the team.

Before the side delivery rake came into general use about 1900, farmers depended upon the sulky or dump rake to gather their hay crop. Furnished with shafts (often called thills), this rake was drawn by one horse and was available in widths ranging from eight to 12 feet.

Beginning in the 1870's, the farmer had his choice of a hand or self-dump machine. The latter version was raised with little effort by pressing a foot lever. The forward movement of the wheels automatically discharged the hay. When the windrow was cleared, the teeth promptly returned to the ground. Some rakes had an adjustment that allowed the teeth to fall to the ground unerringly whether using a slow or fast-walking horse.

Above: Hayloaders designed especially for windrows were built in two styles: single cylinder and double cylinder. The latter style had a cylinder for gathering the hay and another for elevating it. The loader in this haying scene of the 20's is a double-cylinder Moline.

Below: The Dain, a typical gearless loader, permitted the handling of hay from either swath or windrow. It employed two sets of rake bars which elevated the crop with long, gentle strokes. Height of delivery was controlled from the wagon by means of an adjustable apron.

Initially there were two types of side delivery rakes — the fork and the cylinder. The fork rake employed teeth resembling those used on hay tedders of the period and, in a reciprocating manner, they kicked the hay to the side and into a windrow. After 1910 the fork-type rake gradually lost favor to the cylinder rake. Most cylinder rakes could be converted easily from side rakes to tedders by changing the direction in which the reel revolved.

Making its debut at about the same time as the side delivery rake was another great labor-saver — the hayloader. Capable of elevating a steady stream of hay nine or 10 feet, this machine eliminated the colossal task of hand pitching. The work was performed as rapidly as a team could haul a wagon back and forth across a field. Two types were available — the cylinder, and the rake bar or gearless. Most farmers raked their hay before loading, and for elevating windrows they liked the cylinder-type implement best. The gearless rake loaded the hay onto the wagon by means of oscillating bars. One of its big advantages was that hay could be loaded directly from the swath in case rain threatened.

At the barn, an unloading apparatus stood ready to carry the hay to the mow. This apparatus usually included either slings or harpoon fork and a "hay car" or carrier running on an overhead track under the barn roof. Old Dobbin supplied the power.

With an arrangement of ropes and pulleys, the horse pulled a slingload or forkful of hay to the top of the barn, then along the track to the mow. A "trip rope," attached to the slings or fork, dangled within reach of the man on the wagon. When the hay reached a spot directly above where he wanted it to fall in the mow, he would give the rope a sudden jerk. Down would come the hay with a whoosh! Two men with pitchforks usually were needed to level off the pile. "Mowing back" hay was a hard, hot job, particularly under summer temperatures and metal barn roofs.

With a corn binder, five to seven acres a day could be cut and bound in convenient-size bundles. This was nearly as much as a man could cut by hand in a week.

A power carrier on this John Deere corn binder dropped the bundles well out of the way of the team for making the next round. A tongue truck relieved the horses of neck weight and made turning easier at the end of the field.

Although temperatures were generally more moderate during corn cutting, it still was one of the most wearisome jobs of the season. With the old-fashioned corn knife, a man seldom could cut and shock more than an acre and a half a day under the most favorable circumstances. If the corn was tall and the yield heavy, one acre was considered a good day's work.

A corn shock could consist of from 64 to 100 hills, but because it required less "carrying in," the smaller size was more popular. The proper time to cut corn was just as the ears began to glaze. This period was limited to a few days each fall if farmers were to get full feeding value from the fodder. It was often difficult to get the crop cut at the right stage of maturity, because fall plowing, wheat sowing and potato digging also had to be done at about the same time.

The Advance-Rumely husker-shredder was a typical corn harvesting machine of its day. It was built in three sizes: four-, six- and eight-roll. With the exception of the wind stacker, each size was identical in design, and differed only in capacity and the power required.

In the early 1890's a machine appeared that could straddle the rows, cut the stalks evenly, and bind them into compact bundles. The corn binder was considered one of the triumphs of the century. Its principle features were a pair of dividers, chains with fingers for carrying the stalks to the rear in an upright position, the knives, and the binding mechanism. With a good three-horse team, from five to seven acres could be cut in a day. The services of two or three men were required to place the bundles into secure shocks.

Before the advent of power machinery, corn husking was a long-drawn-out, sore-handed job. A good worker seldom could hand-husk more than 75 to 100 bushels a day. First, a shock had to be "torn down." Then while resting on his knees, the farmer stripped each ear of its husk with the aid of a hand "husking peg." The occasional appearance of a red ear — not uncommon with open pollinated varieties — usually served to break the monotony.

As evening approached, the day's yield was gathered into baskets, dumped into a wagon and removed to a crib for unloading. In addition, the fodder had to be hauled to the barn or placed in large shocks for future hauling.

Mechanical corn huskers had been a topic of conversation for years. The first machines, generally owned by threshermen who operated them with traction engines, did the job after a fashion. But they were not entirely satisfactory. About 1900 a machine became available that met the farmer's requirements fully; it not only husked the corn but shredded the fodder as well. The bundles were fed into the machine by hand from a feeding table, in much the same manner as was usual with early grain separators. After passing through the snapping and husking rolls, the golden ears were elevated in an endless stream to a waiting wagon. The fodder, properly shredded for feed and bedding, was blown into the barn loft by a wind stacker. Husker-shredders usually were manufactured with either four, six, or eight husking rolls. The eight-roll size was used mostly for custom work. Under favorable conditions an eight-roll machine could husk from 700 to 900 bushels a day.

This fall harvesting scene was photographed near Batesville, Indiana, in 1916. The corn husker is a four-roll Rosenthal. The tractor is a Big Bull.

The decade of the 1930's was a transitional period in American agriculture. While some machines were being redesigned for tractor adoption, others were falling into disuse and rapidly becoming obsolete. The walking plow, riding plow, grain binder, corn binder, corn husker, and many other implements were performing their last acts on the agrarian stage. They were becoming victims of progress — that eternal quest for efficiency that pervades all fields of endeavor.

But progress is seldom achieved without a price. Traditionally, farming had always been a community activity. Certain seasonal operations could be accomplished effectively only in a reciprocal manner — by neighbor helping neighbor. This practice perpetuated a bond of fellowship that was acknowledged as one of a community's greatest assets. By the late 1930's the trend was toward a new breed of machines that would do away with this interdependence forever.

The early-winter scene of the 1920's shown here features an International Harvester outfit. Large capacity eight and ten roll huskers of the period required the power of tractors such as this McCormick-Deering 15-30.

Above: "Belting up" the threshing rig. The steam plant was frequently as far as 60 feet from the separator. Alignment had to be just right or the belt would jump the pulley. A rope drive was used on this 1899 Ontario rig (Ontario MAF); Below: Sometimes the fireman could be coaxed to "let her smoke" for a photograph. Actually, a good fireman ran clear smoke. Kearney, Nebraska, 1910 (Nebraska SHSM).

2
The Glorious Days of Steam Threshing

A steam traction engine held an insatiable fascination for the boy on the farm. What boy had not transformed his toy wagon into a traction engine by setting an old five-gallon kerosene can in it and rigging up a lot of gears and wheels from a discarded binder? A few corn cobs burning in the bottom of the can completed the realism.

The youngster would thrill with emotion when he heard the faint, clear whistle of a threshing engine far off on some distant farm. He would rush excitedly to inform others when the first black smoke rose above the tree tops far down the road. He would trudge admiringly beside the engine as it lumbered along, ironing the dust into smooth patterns with its wide, cleated wheels.

And how he idolized the engineer! When told "Fetch me that monkey wrench, lad," he would obey with a bored, nonchalant manner, being careful not to show his delight. He even learned to spit like the engineer, although he labored under the disadvantage of not using tobacco.

The smoke-belching steam traction engine was a familiar sight throughout rural America for 50 years. As the prime motive power, its influence greatly affected farm life in general and threshing in particular. Emerging from its embryonic stages in the 1880's, the traction engine ushered in a new era of power farming that freed the thresherman from his dependence upon horses. But his emancipation had been many years in coming.

In the early 19th century, threshing was the most time-consuming and toilsome job on the farm. The well-timed strokes of the flail, although peculiarly adapted to the development of arm muscles, became excessively fatiguing when one followed it day after day for weeks or months. Eight bushels of grain was considered a good day's work for a man.

Threshing completed, the equally laborious task of winnowing had to be done. This was the process of cleaning the grain by casting it into the air, a shovelful at a time, trusting the wind to blow the chaff away. It was practically a winter-long job for a lone farmer to thresh and clean the crop of a 10-acre field. After the Pitts Brothers patented the first successful threshing machine in 1837, hand threshing methods gradually disappeared. And a new role was created for the farm horse.

FEARLESS TWO-HORSE POWER AND THRESHER & CLEANER,

MOUNTED, IN OPERATION.

A typical threshing operation of the mid to late 19th century is shown in this old engraving. With a tread power, it was possible to thresh 300 bushels of wheat a day. The manufacturer of this outfit was the Empire Agricultural Works of Cobleskill, New York, a firm founded in the 1840's.

The first threshing machines were operated by tread powers, which took advantage of the horse's weight. An inclined platform, called the bridge, consisted of an endless chain of planks that revolved as the animals stepped forward and upward. Manufactured in sizes to accommodate up to three horses (oxen were sometimes used), tread powers could thresh from 200 to 300 bushels a day.

Sweep powers gave even greater capacity to threshing machines. These units were adapted for as many as 14 horses, moving at a speed of two and one-quarter miles an hour in a circular fashion resembling a merry-go-round. The horses were controlled by a "driver" who stood on a platform at the center of the unit. Power was transferred to the threshing machine through a bull-gear, spur-gear, and a set of shafts called tumbling rods. It was necessary for the horses to step over this shaft each time around.

This engraving illustrates a sweep power of the period driven by 12 horses. Since it was impossible to use reins with these animals, the driver had to manage them by tone of voice. Note the tumbling rod at left. (J. I. Case Co.)

Almost every manufacturer of threshing machinery of-
fered a line of sweep powers, usually of the Dingee-
Woodbury pattern. Daniel Woodbury of Rochester,
New York, was the inventor, and the name of W. W.
Dingee was associated through his improvements.

Portable steam engines were the primary source of
power for threshing in the 1870's. Cost of the average
unit usually was about $1,200. Like this M. Rumely,
most portable engines were equipped with stacks which
could be folded back when the engines were moved.

Although satisfactory work usually resulted with horse
powers, these rigs occasionally were troublesome. With
tread powers, horses not thoroughly broken to the work
would sometimes "crowd" other horses in the unit.
Sweep powers were difficult to get into motion from a
dead standstill, often causing the horses to rear and
lunge. Horses frequently walked too fast or too slow to
provide proper grain separation. But eventually the
greatest problem became that of power; the maximum
output of the largest sweep power could not meet the demand for increased capacity.

By 1870 the horse was being seriously challenged by a
new entry into the field of farm power. Threshing with
steam, a strange concept a decade earlier, was providing
increased capacity at a lower cost of operation. Portable
engines, with their attendant smoke, sparks and escaping steam, were ushering a new epoch into American
agriculture.

Operating at 150 rpm and controlled by a governor, the average engine was rated at 10 horsepower. Boilers carried a maximum pressure of 90 pounds. One of the unique features of portable engines was their tall smokestack, designed to improve draft. Many were hinged at the base and folded down when the engine was not in use. While enroute to the next job, the driver often controlled the team from atop the engine — in a seat mounted on the stack.

Above: This engraving taken from a Russell and Company catalog of the early 1880's shows a typical threshing outfit of the period on the road. Although the engine was self-propelled, a team of horses was needed to do the steering. The separator was a Massillon "Cyclone," also manufactured by Russell. (Dale Fasnacht)

Below: Grain separators used in the 1870's were of the apron type, having an endless canvas belt to which was attached a series of wooden cups. The example pictured was manufactured by the A. W. Stevens Company of Auburn, New York.

The typical steam outfit in the 1870's could thresh about 800 bushels a day. Threshing machines featured a feeding table, a cylinder, and a canvas belt with a series of cups attached, called the apron. When conveyed to the top of this revolving belt, the grain and chaff dropped onto the sieves below, while the straw was carried from the machine by a raddle.

The grain was delivered through an opening near one of the rear wheels of the separator into half-bushel or bushel measures. It was customary to give "big measure," heaping the container to make up for any light kernels. The straw continued outward, reaching a stack by means of an endless conveyor stacker, which was made adjustable, up or down, by rope and windlass. In later years these stackers were made more versatile, swinging freely from side to side and folding down when not in use.

Portable steam engines eliminated the need for horses during the threshing operation, but in moving from farm to farm they were as necessary as before. Each unit of the threshing outfit had to be drawn individually, and about eight animals usually were required. Custom men yearned for an engine with the tractive mechanism to propel itself.

Left: In addition to being self-propelled, the Case steam engine of 1886 was also self steered. With the availability of units such as this, the thresherman was freed from his dependence upon horses. (J. I. Case Co.)

Below: By 1890, the popularity of traction engines for operating threshing machines was well-established. The Rumely engine of that year, built in La Porte, Indiana, was available in 8-, 10- and 12-horsepower sizes. Drive wheels were rear mounted.

The first steam tractors appeared in the late 1870's. They were frequently called "horse-steering" engines. Although self-propelled, they were not yet self-steering and required the services of a team heading the procession of traction engine, water tender and separator. Self-steering devices became available in the early 1880's, and within a few years nearly all engines were so equipped. Thus, the thresherman could complete his entire operation without the use of horses.

The years that followed witnessed an ever-increasing growth in agricultural steam engines, both in numbers and in horsepower. In fact, they experienced an unprecedented boom that extended until about 1915. From 3,000 traction engines in 1890, annual production increased to 5,000 by 1900, and this figure nearly doubled before the boom declined. Engines of 20 and 25 horsepower were common in 1900, but still larger sizes were to come. By 1910 several manufacturers were producing plowing engines rated at 40 horsepower.

In the early years, opinions differed on how best to transfer power from the engine to the drive wheels. A few manufacturers used a chain, running from the mainshaft to a drive sprocket. Others devised a shaft and a set of bevel gears. But the most satisfactory method, and the one used almost universally by 1900, employed a series of spur gears.

Nearly all traction engines had horizontal-tube boilers. However, there were two general types of flues — the direct-flue and the return-flue. Proponents of each type claimed "advantages" that incited arguments in threshing circles for decades. With the direct-flue or locomotive-type boiler, smoke and heat from the firebox passed forward through several small flues to the smokebox, thence out the smokestack. Proponents declared that this design was stronger, and therefore provided a greater degree of safety, because it was less susceptible to external pressure. The stack, located at the front of the engine, caused less discomfort to the engineer in hot weather.

To add a special touch to the advertising, a few manufacturers gave names to their traction engines which denoted exceptional power and performance. Robinson & Company of Richmond, Indiana, called its engine the "Conqueror." Example shown was built in 1902.

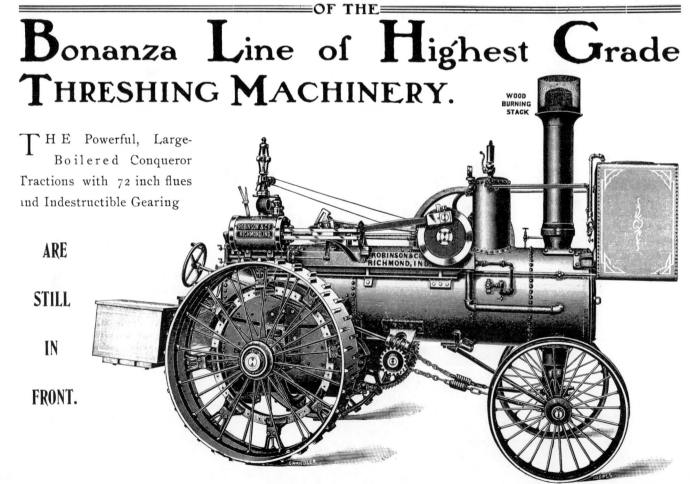

NOW READY. 1902 FREE CATALOGUE
OF THE
Bonanza Line of Highest Grade
Threshing Machinery.

WOOD
BURNING
STACK

THE Powerful, Large-Boilered Conqueror Tractions with 72 inch flues and Indestructible Gearing

ARE

STILL

IN

FRONT.

ROBINSON & Co
RICHMOND, IND.

ROBINSON & Co.
RICHMOND, IND.

The Rumely separator of 1890 was a transition between the old-style apron thresher and the 20th century vibrating thresher. An elevator was available with this machine for loading grain into wagons.

The Huber Manufacturing Company of Marion, Ohio, was the foremost producer of traction engines of the return-flue type. The company claimed this type engine produced a greater percentage of heat with a given amount of water and fuel. Pictured is an 18-horsepower Huber of 1903.

The Aultman & Taylor Machinery Company of Mansfield, Ohio, entered the threshing business in 1866. For years its bevel-gear traction engine was one of the most notable of this design on the market. The thresherman had his choice of either simple or compound cylinders on this 1904 model.

This crew is minting the golden grain near Prescott, Iowa, in 1906. The Gaar-Scott outfit consisted of a 13-horsepower engine and 33x52 separator.

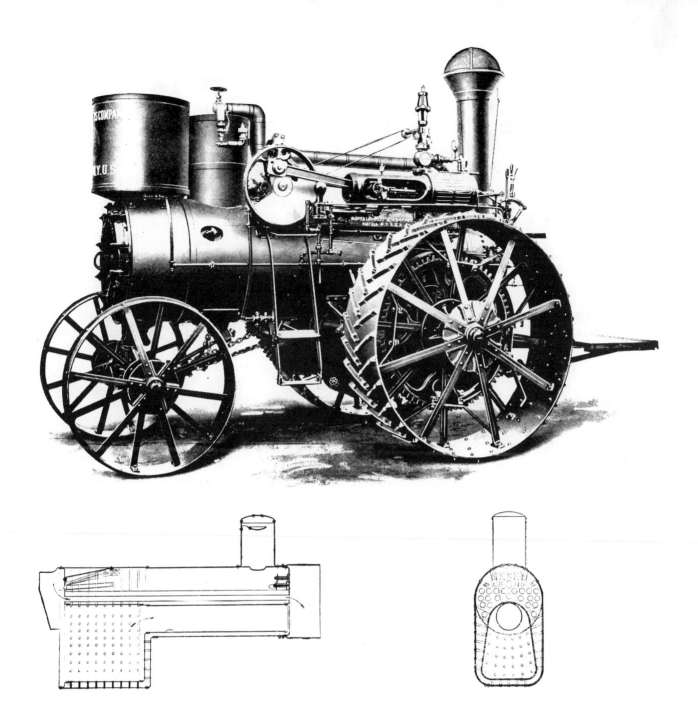

With the return-flue boiler, combustion products passed forward through a large central flue, then upward, returning again to the rear through smaller flues. Advocates claimed more efficient engine operation because the heat passed twice through the length of the boiler. Also, with the rear stack, smoke was not blown into the engineer's face when running against the wind.

Opinions differed also concerning the distribution of weight. The wheelbase of a traction engine varied according to how the drive wheels were mounted. On "rear-mounted" engines, the main axle was held in position at the rear of the boiler, just below the fire door, providing a long wheelbase to "overcome the objection of rearing in front." According to many observers, this method also afforded greater strength.

The Pitts brothers, Hiram and John, patented the first successful grain threshing machine in 1837. Their inventiveness resulted in the formation of the Buffalo Pitts Company. Buffalo Pitts was one of several firms that built both direct and return-flue engines. This machine is a 25-horsepower return-flue model of 1907.

Equally popular was the "side-mounted" engine, whose traction wheels were supported by stub axles, attached to the sides of the firebox. By receiving more weight on its drivers, this design was said to provide better traction and easier steering. Then, too, the shorter wheelbase reduced the area required to turn.

In addition to return-flue and conventional direct-flue traction engines, the Avery Company of Peoria, Illinois, built a double-cylinder "undermounted" engine. The cylinders and other working parts were located on a steel framework below, and independent of, the boiler. Pictured is a 30-horsepower undermounted Avery of 1908.

Hauling water to supply the traction engine was an important job at threshing time. This Avery water wagon had a capacity of 475 gallons, or a little more than 15 barrels.

Agricultural steam engines generally fell into two classifications — the "simple" engine and the "compound" engine. Most tractioneers preferred the simple engine. With it the steam was admitted to, expanded in, and exhausted from, a single cylinder. The compound engine, also popular, featured two cylinders placed end to end, with both pistons driven by the same connecting rod. This engine expanded the steam twice before exhausting, thus theoretically giving greater economy in use of fuel and water. However, to realize maximum efficiency a higher initial steam pressure was necessary, and unless conditions were favorable the savings over a simple engine were negligible.

In an era when high speed was usually associated with railroad traffic, accidents on public highways were, paradoxically, frequent occurrences. Many such accidents were caused by the prevailing disposition of a horse. A spirited horse, chafing at the bit, could be difficult to control with the approach of a steam traction engine. The following account, appearing in an 1897 issue of the **Ashland Press**, Ashland, Ohio, is typical of what could happen:

The Geiser Manufacturing Company of Waynesboro, Pennsylvania, was one of the most prominent builders of threshing machinery in the East. Its "Peerless" tractor engines were well known wherever grain was grown. This Peerless general-purpose engine is a 1912 model.

Aaron Emick had the misfortune of having a runaway last week. Monday evening, when returning home from town, he met Crone Brothers' sawing outfit which was just coming out from Henry Riley's farm. The traction engine came to the road just about the same time Mr. Emick did, and the horse became frightened, wheeled about and started towards town. Mr. Emick was thrown out and a barrel of apples and other articles in the buggy were scattered along the road. The horse and buggy went on to town and collided with an electric light post; the buggy was torn from the horse and badly damaged . . .

This 32-inch Case typifies the general design of the 20th century threshing machine. The self-feeding, wind-stacking, and automatic grain-registering attachments were a boon to farmers and threshermen.

The J. I. Case Threshing Machine Company of Racine, Wisconsin, was the nation's leading builder of both steam engines and separators. Its record of producing more than 35,000 traction engines was unapproached by any other manufacturer. This 50-horsepower Case engine was a popular size because of its wide adaptability. NOTE — Case horsepower ratings differed from those of other companies.

Those who operated steam traction engines were often subjected to annoyance, and even prosecution, regarding use of public highways. Laws arbitrarily regulated highway travel according to the whims and fancies of the legislators. In 1900, an editorial in the **Threshermen's Review** observed:

The average state congressman knows about as much of running a traction engine as a hog does of teaching Sunday school, and when called upon to vote on laws governing their use, his total knowledge is probably bound up in the fact that he once met one on the road and was more scared than the horse he was driving.

Typifying these laws are the following which appeared on the statutes of various states in 1899:

— Stop 100 yards away on meeting a team, or more if the horses are frightened.
— Drive to one side of the road and stop when you meet a team and don't blow your whistle while it is passing.
— Send a man at least 50 yards ahead of the engine to warn drivers of teams.
— At night send a man with a red light 20 to 30 rods in advance of the engine to signal the engineer when horses are approaching.

With the obvious intent of providing friendly assistance to the drivers of horse-drawn vehicles, these laws nevertheless were ridiculous in the eyes of engine men.

There was hardly a major agricultural state that did not initiate some form of anti-engine legislation at one time or another during the threshing era. In their zeal for good roads, a few states placed restrictions upon engine travel because of the injurious effects of wheel lugs.

Above: In 1914, Nichols & Shepard manufactured its single-cylinder traction engine in 13-, 16-, 20-, 25- and 30-horsepower sizes. The separator built by this company, one of the more popular in threshing circles, was called the "Red River Special."

Below: A Case 50-horsepower engine and 32-inch separator made a "dandy" threshing outfit. In this scene, a Case touring car was thrown in for good measure.

In 1913 the Pennsylvania Highway Commission established some drastic rules which made it almost impossible to operate traction engines on public roads. In addition to specifying the kind of wheels that would be allowed, it also instituted a heavy license fee for all operators. The commission became so formidable that farmers and threshermen organized and demanded their rights. The Traction Engine Bill, which passed the Pennsylvania legislature in 1915, was the result, giving these machines the right to use highways the same as any other vehicle.

A new principle in thresher design was generally adopted during the 1890's. Developed by the J. I. Case Threshing Machine Company, it was a complete departure from the apron machine of the period in that it employed racks which shook the grain from the straw and chaff. Most of the features introduced at this time were retained in threshing machines built thereafter. J. I. Case also introduced the all-steel threshing machine in 1904.

Prospective buyers of traction engines built by the Keck-Gonnerman Company of Mount Vernon, Indiana, had their choice of either single or double cylinders. This single-cylinder engine of 1914 was rated at 20 horsepower.

Although some threshermen used grain separators for harvesting clover seed, a specially-constructed machine called the clover huller did a much better job. The Birdsell Manufacturing Company of South Bend, Indiana, built a huller that was world renowned. A rasp hulling cylinder "rubbed out" the seed.

In the early days, the man at the feeding table of the threshing machine was the most important figure in the threshing crew. Feeding was an art. The bundles were severed by the "band cutter," and passed on to the "feeder," who spread them out and fed them, heads first, into the machine. He did this in such a way that the stalks were fed gradually, from the top of the bundle downward. In this manner, the grain flowed into the cylinder in a steady, uniform stream. This was essential for doing a fast, clean job of threshing.

Traction engines manufactured by the A. D. Baker Company of Swanton, Ohio, featured a cylinder of unusual design, called the "uniflow." This cylinder was said to reduce the amount of fuel required to operate the engine. A Baker traction engine of 1916 is shown.

At the turn of the century, a new attachment came into general use that eliminated the dusty and tedious job of hand feeding. The self-feeder offered decided advantages. Essentially, it was made up of a bundle carrier, a band cutter, and a cylinder feeding apparatus. The latter worked on much the same principle as was used in hand feeding. As the grain was pitched onto the slatted carrier raddle, it was conveyed forward, passing under the whirling knives of the band cutter. Then, a retarding mechanism held back the under part of each bundle, while the cylinder was fed from the top. If the carrier became overloaded, its motion was checked by a governor, automatically regulating the volume of grain that entered the machine. This helped prevent choking or "slugging" the cylinder as a result of careless bundle pitching.

These men are "hard at it" on a late-summer day in 1915. There were mighty few breathing spells for a threshing crew.

The Port Huron "Longfellow" was one of the more popular traction engines using compound cylinders. The drive wheels of this engine were equipped with corrugated rims, said to be self-cleaning and non-skidding. This 19-horse Port Huron was built in 1916.

During World War I, the Advance-Rumely Thresher Company introduced a new "Universal" traction engine which incorporated many features of merit found in earlier models. It proved to be very popular. Sizes available were 18-, 20- and 25-horsepower.

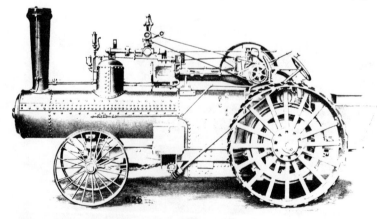

The Frick Company of Waynesboro, Pennsylvania, was one of the pioneer manufacturers of agricultural steam engines. Traction engines offered by this company possessed an independent steel frame, extending from front to rear axle, upon which the steam boiler was mounted. Pictured is a Frick "Eclipse" engine of 1921.

Appearing about the same time as the self-feeder was the wind stacker, often called the "blower." It soon became an indispensable appendage of the threshing machine. Capable of shooting a cloud of straw and chaff about 35 feet into the air, it banished the disagreeable work of hand stacking and saved the wages and board of several men.

The stacker chute telescoped so that it would extend to almost twice its folded length — about 25 feet. By means of an automatic oscillating device, the chute could be made to swing over any desired arc of a circle without the attention of an operator, or it could be manipulated by hand to assist the men on the stack. The deflecting hood was controlled by ropes and had a universal motion — up, or down, or sideways. This permitted the straw to be delivered to almost any location.

Despite the sweeping success of the pneumatic stacker, a few farmers insisted that the wind from the fan sucked the grain from the separator sieves and deposited it in the straw stack. Their suspicions were proven groundless. However, any negligible grain loss was emphasized resoundingly as the kernels were blown against the metal deflecting hood. Noise was, in fact, a component part of the wind stacker. On a still day, a threshing operation was audible to everyone in the neighborhood.

A well-proportioned, rain-resistant straw stack was considered a work of art. Its designer, the man or men who kept the straw evenly distributed and well tramped, had to anticipate its proper diameter at the ground level, gradually draw in its sides as it extended upwards, and finally see that its peak was nicely "topped out" when the last bit of straw left the separator.

In the late 1880's some threshers were equipped with an elevator for loading the grain into wagons. Many also had a bagging attachment and a tallying device for counting the bags. These attachments were mounted to the separator in a manner which allowed a limited choice of positions for receiving the grain. But this began to change in the late 1890's. Swinging conveyors with swiveling wagon spouts permitted delivery to any convenient point at either side of the machine. Attachable bagging spouts could be placed in innumerable positions. Complementing this new grain-handling machinery was an automatic registering device that tallied the grain by weight.

With the addition of the self-feeder, wind stacker and automatic weigher, the grain separator was transformed into a truly efficient machine of greatly-increased capacity. These attachments reduced the manpower requirements for operating a threshing machine by nearly 50 percent.

The size of a threshing machine was indicated by the length of the cylinder. Although 30- and 32-inch sizes were quite popular in the east, many machines were manufactured with 36-, 40-, and even 44-inch cylinders during the heyday of steam power, providing enormous capacity for the largest operations. A 40-inch machine could thresh up to 3,500 bushels of wheat in a 10-hour day.

At no time of year did men on the farm work at such high tension as they did at threshing time. It was a period that demanded total involvement and hustle at top speed. Help was exchanged throughout the neighborhood, as farmers worked together to keep the huge rig running at full capacity. There was always the possibility of rain, and every effort had to be made to rush the job along while the weather was favorable.

While the threshing crew labored amidst a cloud of dust and chaff, the women folks were busy in the kitchen making ready the best dinner that variety, skill, and competitive spirit could prepare. The piercing shriek of the steam whistle and the sound of the thresher slowing to a stop were signals to head for the dinner table. But first, the crew stopped at a wash bench supplied with generous quantities of water, soap, and towels.

Days before the threshers were scheduled to arrive, the farmer's wife started to plan the gargantuan meals she was going to serve — huge batches of home-baked bread, ham, beef, mashed potatoes galore, vegetables, gallons of coffee, several kinds of pie. At such times chickens died in numbers from the violence of the farmer's axe, and the post-mortem of the remains took place at the threshing table in the presence of a jury of hungry men. The excellence and great variety of the bill of fare usually tempted the crewmen to eat more than was good for them. The expression, "to eat like a bunch of threshers," still lives on in our lexicon.

Custom threshing was an institution in rural America for several decades. To many a young boy, the lot of the thresherman was the pinnacle of ambition. To each farmer, the annual arrival of the custom rig represented the culmination of a year's toil and hope. During the heyday of steam power only one farmer out of every 20 owned or operated a traction engine, and upon him the remainder was dependent. With their ranks once estimated at 75,000, threshermen exerted a profound influence on rural life.

Characterized by grease-besmirched clothes and a chew of tobacco, threshermen were a dedicated aggregation of individuals. While a few were motivated by the desire for public esteem, the majority considered their work a way of life. Many had grown up around a steam engine and had threshing in their blood.

Although the difficulties encountered in custom threshing were many, and the financial rewards modest at best, threshermen seldom retired. The steady hum of the threshing cylinder, the bark of the engine's exhaust, the matchless food served at mealtime, the general satisfaction of a job well done — all lured them back year after year.

Although the application of steam power to grain threshing afforded unquestionable advantages over horse-power methods, it also introduced hazards that occasionally marred the threshing scene. The anxieties induced by potential dangers were inescapable.

Despite all precautions, sparks discharged from the engine smokestack were a constant threat. Straw, a highly combustible material, was usually nearby and

Above: Russell & Company of Massillon, Ohio, began
the manufacture of threshing machinery in 1842. The
company is credited with building the first practical
friction clutch for use on traction engines. This Russell
engine of 1923 was equipped with extra side tanks,
used when additional water was needed for long hauls.

Below: The Harrison Machine Works of Belleville, Illinois, manufactured a traction engine called the
"Jumbo." This engine in 1923 was manufactured in
15-, 17- and 20-horsepower sizes.

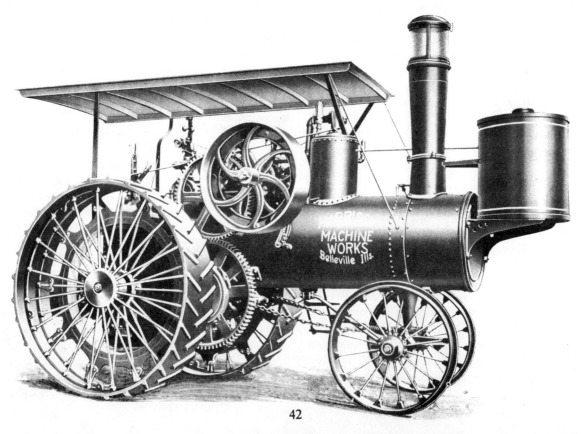

added to the danger. Farmers sometimes disposed of old fence rails at threshing time by using them as fuel for the engine. Firing an engine with wood was particularly dangerous, as this tended to create a shower of sparks, and occasionally a disastrous fire. Frequent fire losses ultimately forced the use of coal in steam engines almost everywhere.

Accounts of fires emanating from steam engines can be found in the early-day newspaper files in any grain-growing community. The **Ashland Press** carried this report in July, 1900:

Friday noon when Geo. Emmens was taking his engine past Wm. Kendig's barn a mile north of town, a spark from the smokestack set the strawstack on fire, and in very short order a large blaze was on hand. The neighbors were aroused by the blowing of the whistle, and soon were there working heroically to save Mr. Kendig's valuable barn. It happened that the wind was strong, but was blowing the right direction, and the building was thereby saved with the additional assistance those present rendered by pulling the straw away from the barn and throwing water on it.

To "deliver the goods" successfully, a steam traction engine required an engineer's almost constant attention. The fire had to be replenished at short intervals and a vigil kept on the water supply at all times. Whether the result of negligence or other factors, a low water level in the boiler was always an alarming discovery. It was necessary to bank the fire quickly, and sometimes to also elevate the front of the engine, allowing the remaining water to gather overtop the firebox. If not done in time, the result could be a collapsed crown sheet or a split flue, temporarily disabling the engine. Or, the consequence could be disastrous. According to the **Ashland Press** in July, 1901:

The boiler of Ora Emmens' threshing engine blew up Monday forenoon with marvelous results and miraculous escapes. The engine and boiler were demolished and parts were hurled hundreds of feet.

Emmens owns the engine and John Kissel the threshing machine. They were threshing at Clint Boyd's, three miles north of Ashland, in the yard just east of the barn. Jay Jackson had charge of the ten-horsepower engine.

The boiler had been leaking and it was brought to Mohn's shop in town last Saturday. Mohn put a plug in the boiler and pronounced it safe. About eight o'clock Monday morning Jackson noticed that the boiler was leaking slightly. He at once notified Kissel, then stepped upon the footboard of the engine, signalled a stop with the whistle, reversed the lever and had just stooped down to scrape out the fire when the explosion occurred.

The noise was deafening and the effect awful. A huge cloud of dirt and steam enveloped everything. With a tremendous force the huge engine and boiler, excepting the one drive wheel, was lifted 20 feet from where it stood while parts were scattered everywhere. Jackson was hurled far and away to the southwest alighting in the field where clover had been cut. The distance was afterward measured and found to be 142 feet. His escape from death was not much less wonderful than that of John Kissel who stood about 12 feet to the rear and left side of the engine. The steam and water escaped toward him, knocking him a distance and only injuring his face some . . . The worst injury came to John Wertman who was feeding the machine. What is supposed to be the belt struck him

upon the head, cutting a gash about six inches long. Pieces of iron just missed him and made dents in the machine as he leaned over for a sheaf . . .

The feature that caused more comment than anything else was how those injured and uninjured escaped as well as they did. Jackson's escape was most remarkable. He attributes it to the fact that he was stooping down. Had he been standing up the flying pieces would have struck him. Shortly after he alighted on the ground he arose and in a dazed condition wandered around in a circle with his hand over his eyes. He had four or five cuts on his face in the vicinity of his left eye, but none of them appeared to be dangerous. His right arm received a terrible bruise near the elbow and swelled to nearly twice its size, but further from those injuries he was unhurt. His escape so lightly makes it doubtful in some minds that he was thrown so far, but those who were there say there was no doubt about it as they found where he struck and measured the distance.

The cause of the explosion cannot be attributed to any fault of the engineer, but to the weakness of the boiler. The leak that had been discovered started the vent, and after the explosion it was discovered that quite a seam was rusted out. Surprise was expressed that the boiler held as long as it did . . .

In 1911 it was estimated an average of two boiler explosions occurred in the United States every day. In the first six months of 1914 there were 340 boiler explosions, 120 people were killed, 240 were injured, and property loss amounted to $250,000.

The crown sheet, which was located directly over the fire, was generally the first part of a boiler to show weakness. It required every possible safeguard against the stress of steam pressure. It became most vulnerable when engines were moving down hill, because the water, tending to seek its own level at the front of the boiler, often exposed the sheet to the fire.

To provide a degree of safety, a fusible plug was inserted in the crown sheet of each engine. If the water became too low, the plug would melt, allowing steam to escape into the firebox and extinguish the fire.

Another serious problem that plagued steam threshermen was the great number of inadequate bridges. Wooden structures that had been built to support horse-drawn vehicles were placed under a tremendous strain by the weight of a 10- or 12-ton traction engine. Threshermen often traveled miles out of their way to avoid bridges that were thought to be unsafe. Occasionally they transported heavy planks which they could place over the floors of questionable structures before crossing. Eventually streams were spanned with steel and concrete, but in many cases not until steam power was passing from the rural scene.

The toll of human lives that resulted from collapsed bridges forms one of the tragic chapters of American agriculture. In August, 1904, the **Ashland Press** reported the following:

As the result of a traction engine going through a bridge Thursday afternoon, William Garn, a young man about twenty-five years of age, died withing twenty-four hours and another, Arthur Zimmerman, about the same age, lies at his home seriously, if not fatally, injured.

This steam traction engine of 1923 was part of the "Great Minneapolis Line" of power farming machinery. It was built in 20-, 24- and 28-horsepower sizes. Platform tank at rear of engine was furnished at extra charge.

The awful accident happened while the young men were steering a traction engine across a wooden bridge leading from Nelson Cameron's to Squire James Cameron's farm. The threshers had finished threshing for Squire Cameron and had gone to Nelson Cameron's. They finished there Thursday afternoon and instead of taking the same road they came on, they took another, a private road, which would save a considerable part of the distance to be traveled. Mr. John Eagle stated Monday afternoon when telling of the dreadful accident that Mr. Cameron had cautioned the threshers not to take the road as he believed the bridge to be unsafe. Instead of taking his advice, Martin Zimmerman, the father of Arthur and one of the county's veteran threshermen, took an axe, went to the bridge, examined it, and pronounced it safe. The run spanned by the bridge is eight or nine feet deep at that point.

The bridge proved in fact wholly unsafe as hardly had the big engine got upon it when it gave way and the engine with the two young men upon it was precipitated to the bottom of the run. Both men were pinioned beneath the engine and were frightfully scalded by the escaping steam. Mr. Eagle who was one of the five men who were present to help rescue the men said their agonizing cries for help were heard by neighbors over half a mile away. The scene was heartrending and the rescuers worked frantically in their efforts to save the doomed men. Young Zimmerman was the first one rescued and it was fully twenty minutes before Garn was taken from beneath the engine more dead than alive. Both men were at once given every medical attention but it was early seen that Garn was fatally injured. He died the following afternoon after hours of the most intense suffering . . .

Accidents at threshing time were not always engine-related. The self-feeder, with its slashing band cutter knives, represented a hazardous area for the inattentive bundle pitcher. Stepping across a main drive belt as it flopped in the wind always presented the element of danger. And every now and then there was the element of surprise when the unexpected happened. The **Ashland Press** related the following incident in October, 1899:

At Henry Long's farm near Hayesville last Wednesday, an exciting accident happened. Moffett Bros. were the threshers and while the engineer was in the barn the belt regulating the governor of the engine came off, giving the threshing machine an uncontrollable and intense speed. The effect was to break the cylinder to pieces, and the flying pieces shattered the front part of the machine and tore holes in the barn roof as big as a man's head. All of the helpers instantly dropped in their places and thereby escaped being hurt or killed. The engineer ran the most risk by running out of the barn and stopping the engine.

Depending on the locality, grain was threshed either directly from the shock or from the barn floor. When shock threshing, the custom man usually commenced his "run" in late July and, weather permitting, moved steadily from farm to farm. From a considerable area around each "setting," farmers with their teams and bundle wagons gathered to assist in the work. When grain was stored in barns to allow it to "sweat" for a period, the threshing operation was sometimes delayed until after the wheat was sowed or the corn cut. But ordinarily the threshing season was over by early autumn.

To increase the use of their engines, some threshermen operated sawmills during the off-season. There was no activity that demanded greater skill from a sawing crew than that which preceded the numerous "barn raisings" of the period. Weeks before the eventful day, the engineer moved his portable sawmill into a timber lot on the farm where the raising was to take place. Presently the crew sawed out the huge beams, sills, and other complex pieces of the barn's framework.

When the stone wall was completed and everything was in readiness, invitations were sent to friends and neighbors. Few persons declined the invitations. Barn raisings were a striking example of the cooperative spirit that once prevailed throughout rural communities.

In one Ashland County community more than 100 people were present in 1898 as a 40 by 100 foot barn was raised by the pike pole and heave-ho method. A few miles away, 45 men and 15 women were on hand as a 36 by 70 foot barn was erected in 1905. Women ably presided over the culinary department on these occasions.

Sawmills were illustrated in the literature of many thresher manufacturers during the steam era. This example appeared in an early Case catalog. (J. I. Case Co.)

Tables laden with a variety of delicacies often "groaned" from the weight and caused the men to groan all afternoon from overeating.

For operating a portable sawmill, an engine of 60 brake horsepower usually was recommended. Even in later years, sawyers often were partial to steam engines for this work. This was generally because slabs were available as convenient and economical fuel, and because many sawyers found that steam power was more flexible.

While the steam traction engine was used primarily for threshing and other stationary belt work, sometimes its use was extended to drawbar functions. There was no greater power requirement in agriculture than that of plowing. The settlement of prairie lands west of the 100th meridian had long been retarded for lack of power to break the miles of stiff virgin sod. To attempt such work with horses represented a Herculean task. It was calculated that a man and team would have to walk more than 5,000 miles to turn over one square mile of land with a 12-inch plow.

Attempts at harnessing the plow to steam power were made in the 1880's, but with questionable success. A string of two-bottom gang plows, hitched in tandem, bobbled their way along, left the ground at the least provocation and posed a problem in turning at the end of

The sawmill in this scene was operated by a Frick 20-horsepower portable engine. The log being sawed was 52 feet long and made a finished beam eight by ten inches in diameter.

the field. Moreover, traction engines of the period were not designed for drawbar work; their drive wheels were too narrow and their gearing too weak.

By 1904 engine gang plows with steam-lift attachments were being introduced. Available in many sizes up to 14 bottoms, these huge plows were suitable for anything from a quarter section to a North Dakota ranch. Manufacturers also began building special plowing engines that offered increased power, strength, and durability.

Above: The Geiser Manufacturing Company was one of the pioneers in building steam plowing outfits. Its plowing engine of 1903 featured a patented "crown sheet protector," which minimized the possibility of boiler explosions when moving downgrade.

Below: The Advance Thresher Company of Battle Creek, Michigan, manufactured an extensive line of traction engines, ranging in size from 10 to 40 horsepower. This 40-horse, cross-compound plowing engine was built as either a coal and wood burner or a straw burner.

Breaking tough Dakota sod was the job assigned to this Advance plowing engine in 1909.

The Reeves 40-horsepower, cross-compound engine of 1913 was built for large-scale plowing operations. The drive wheels were regularly equipped with 28-inch rims, but these could be widened to 56 inches by adding special extensions.

The Wood Brothers 30-horsepower plowing engine of 1914 was unique in that it was double-geared. Distributing the engine's power directly to each drive wheel gave a straight, balanced pull and eliminated the strain and friction sometimes experienced with single-geared machines.

The Gaar-Scott "Big Forty" plowing engine had double-compound cylinders. It came equipped with an extra large coal bunker, one large rear tank and two side tanks, with a total water-carrying capacity of 21 barrels. Weight was 35,000 pounds.

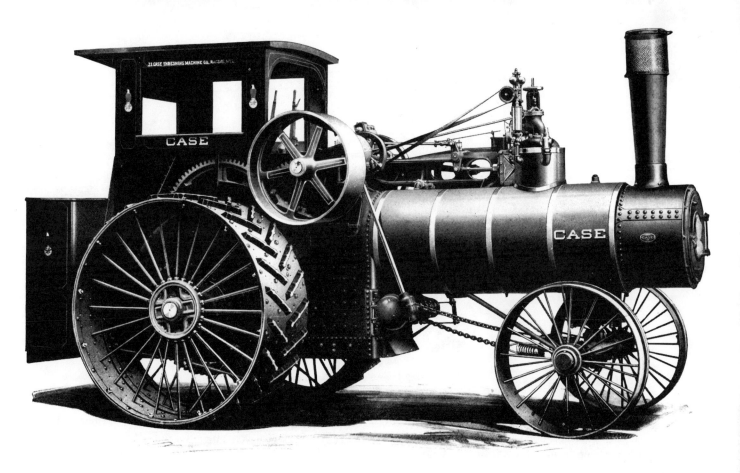

This 110-horsepower giant was the largest-size engine manufactured by J. I. Case. In 1912 and 1913, these engines won gold medals in their class in the annual plowing contests held at Winnipeg, Canada. Price in 1915 — $3,000.

Threshing is done with a small individual separator in this scene near Dixon, Illinois, in 1920. A Fordson tractor furnishes the power.

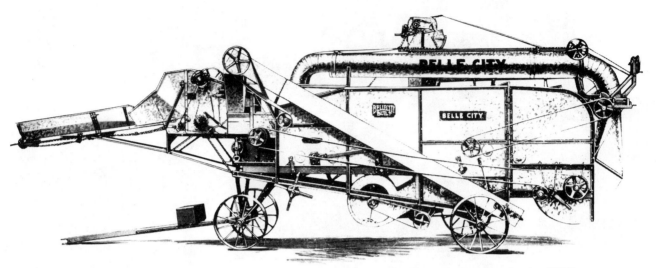

Small "individual" grain threshers such as this 24-inch Belle City became quite popular in the 1920's.

Semi-steel or all-steel gears replaced those of cast iron. Some engines were equipped with double gearing, which distributed power from the crankshaft directly to each drive wheel. The axle, mainshaft, and countershaft were strengthened. To improve traction, wide drive wheels with optional extension rims were provided. Larger water tanks and coal bunkers permitted longer runs without replenishing. To these features some manufacturers added double cylinders. This was said to insure smoother engine motion with no shock to shafts or gears while starting.

By 1910 many steam-plowing engines had a rating of from 30 to 40 horsepower. Engine weight ran as much as 16 tons. Drive wheels from six to seven feet in diameter and 30 to 36 inches in width were common. Boilers made of steel, with double riveting at points of greatest strain, safely maintained steam pressures of 160 pounds.

Two men, operating an engine and 12-bottom plow, could turn over more than 30 acres in a day, replacing six men and 24 horses doing the same job. The services of two additional men, with teams, was usually required to keep the engine supplied with fuel and water.

An important part of the field equipment for steam plowing was the repair wagon. It was equipped with a work bench, boxes for storing tools and materials, an anvil, a drill and a vise. When anything went wrong with the engine, the repair wagon was hauled to the scene of the problem and the work completed promptly.

By 1917 the trend toward light-weight gasoline and kerosene-burning tractors was creating a demand for smaller threshing units. The manufacture of 22- and 24-inch "individual" machines became general, since these were ideal for use with a two-plow tractor. Because bad weather and breakdown often made it uncertain when the custom rig would arrive each year, a number of farmers purchased these new machines. By owning a

small outfit designed for individual use, a farmer could thresh at his own convenience, and often with just his own crew, or at most, the help of a few neighboring farmers.

But the 28-inch threshing machine became the favorite of those wishing to do away with the annoyance of waiting on the custom rig. It could separate up to 1,200 bushels of wheat in a 10-hour day, and its power requirements were moderate. This size found wide acceptance in the community threshing "rings" formed after World War I.

A threshing ring was an arrangement whereby a group of farmers shared equally the initial cost of a rig, then worked as a unit annually until each member's grain was threshed. So one by one the larger machines were discontinued until, by 1930, no thresher with a cylinder exceeding 32-inches was manufactured. This size remained standard for custom work until the threshing era came to a close.

Coinciding with the advent of the small grain thresher, operated with the ordinary farm tractor, was a marked increase in the use of larger tractors for custom work. Many men who had been prejudiced in favor of steam were becoming enthusiastic boosters of gas rigs. While the investment in a tractor was usually smaller, the utility was greater, because it could be used the year around for either belt or drawbar work. In contrast, the steam engine usually stood idle for many months each year. The lighter gasoline tractor could cross bridges that had to be planked for the steam engine — or avoided altogether. The gas rig could move and set up at the next job in less time; there was no waiting to get up steam in the morning, or delays to take on fuel or water enroute. Then, too, a tractor-driven outfit could be operated with a smaller machine crew.

This Ohio crew nears the end of a busy day's threshing in the early 1930's. The tractor is a Huber 32-45.

As the popularity of steam power steadily declined during World War I, it became obvious that an era was drawing to a close. Traction engine sales slumped to 1,700 units in 1920, and only the most popular sizes continued to be manufactured. By 1925 production virtually ceased. The familiar blasts of the steam threshing whistle, echoing across the countryside, were being consigned to oblivion.

The hum of the threshing machine, powered by gas tractors, continued to dominate harvesting activities — but for only a brief time. As combine sales began to soar in the late 1930's, it was clear that the excitement and intensity of the annual threshing scene would soon be just a cherished memory.

OilPull tractors were sold by the Allis-Chalmers Manufacturing Company for a few years during the 1930's. A 30-50 A-C Rumely OilPull supplies the power in this threshing scene of 1933.

Specifications of Case 32x54 Thresher

CONSTRUCTION—Structural steel frame. Sides and deck of galvanized sheet steel securely riveted to frame by powerful air hammers.

TRUCKS—Built-up wheels 34″ in diameter; 8″ steel tires, regular (10″ on order at extra cost). 20 spokes with countersunk heads. Lock nuts on spokes at hub. Axles, two 5″ steel channels with 12″ skeins securely bolted.

HITCH—Short tongue for tractor. Slip tongue for horses furnished at extra cost.

CYLINDER—32″ in length, 32″ in diameter. 155 steel teeth with hardened blades and annealed shanks, interchangeable with concave teeth. Guards prevent wrapping at cylinder boxes. Speed of cylinder 750 R. P. M.

BEARINGS—Roller bearings on cylinder shaft and wind-stacker fan shaft. Bearings on crank shaft, beater and fan shaft are of the ball and socket, self-aligning type, babbitted and fitted with shims to take up wear.

CONCAVES—Three two-row concaves and two open hearth annealed blanks are regularly furnished. Concave circles and steel grate back of concaves are both adjustable.

Extra Profits from Custom Threshing

BEATER—Has four concave sheet steel wings and runs in same direction as the cylinder. The beater will not wrap nor wind. Driven direct from cylinder.

STRAW RACK—Wood, open slat work with five risers. Two main rails supported by four maple boxes (boiled in oil), protected by metal sand shields. Separating surface, 70½ sq. ft. 230 vibrations per minute. 3¾″ throw. Throw of crank, 7″.

GRAIN PAN—Wood side rails and cross pieces; galvanized sheet steel bottom with longitudinal spreading ribs which distribute grain evenly on the pan. 2⅜″ throw. 230 vibrations per minute.

CLEANING DEVICE—Under-blast fan 28¼″ in diameter with four blades. Speed 485 R.P.M. Blast regulated by adjustable upper and lower fan blind on each side of fan drum. End shake shoe. Sieves regularly furnished: one adjustable chaffer with adjustable extension, one adjustable shoe sieve. One 1/14″ x ½″ cheat screen or alfalfa sieve or timothy sieve if stated in order. About thirty other kinds of non-adjustable sieves are carried in stock for these machines, any one of which may be had on special order at small extra cost.

TAILINGS ELEVATOR—Steel double-tube type. Means of elevation—steel flights and steel sprocket chain. Drive to upper shaft by belt from crankshaft.

INTERIOR IN GENERAL—Width between rear posts, 54″. Average height, shakers to deck, 26¼″. Adjustable metal flap behind beater deflects grain downward; canvas flap behind metal flap stops flying kernels.

GENERAL DIMENSIONS—Wheel base, 151⅞″. Length over all, thresher only, 17′ 1″; with feeder and windstacker attached, 27′ 9″. Extreme width, 7′ 9¼″. Height from ground to top of deck, 8′. Height over all (top of hood of windstacker pipe resting on its support) 10′ 4″.

ATTACHMENTS ON SPECIAL ORDER — Windstacker, self-feeder, grain handler, brake, clover and alfalfa attachment, clover or grain recleaners, rice, pea and bean attachments, speed reducing countershaft.

MAIN BELT PULLEY—To furnish the proper size of main cylinder pulley, we must know the diameter and speed of the belt pulley on tractor or engine to be used. Without this information, no pulley will be shipped.

3
Those
Hit and Miss
"Hired Hands"

It was a remarkable sight in 1890 to see a gasoline engine in action, operating automatically without the aid of a great water boiler, a huge bulk of coal or wood to keep it hot, and an engineer to constantly attend it. Like many prototypes in new fields of manufacture, it seemed strange in the general order of things. However, in one respect it differed little from the steam engine — its weight was ponderous. An engine of two horsepower often weighed as much as 4,000 pounds. The cylinder was set upon a massive cast iron base some two and one-half to three feet high, and it had flywheels that measured up to five feet in diameter. All this was considered necessary to properly accommodate a two-horsepower outfit.

The degree of uncertainty connected with the operation of these engines often confounded even the most proficient steam engineer. A man with little mechanical experience usually was unable to get one to run at all. But as crude and unreliable as they were, constant improvements during the ensuing years gradually brought them nearer to perfection.

Back at the time of the Civil War, horses performed the heavy power requirements of the farm. Even then tread and sweep power rigs were scarce. Only the more important work was done by power. Then came the steam engine, efficient in its day, but not at all adapted to the individual farmer's needs. Its successful operation required too much skill, and its cost was too great.

By 1900 a convenient and economical source of power was availing itself to the farmer, enabling him to handle more work in spite of a diminishing supply of farm labor. Wages were increasing too, clearly indicating that to succeed he must take advantage of labor-saving devices. The gasoline engine was beginning to fill a niche in the economy of the American farm.

Progress was characterized by the adoption of the gasoline engine, often dubbed "the hired hand," to farm tasks that had previously been back-breaking chores. In its earliest stages, the gas engine performed very simple duties. It could not be harnessed to a variety of machines, because implement manufacturers had not designed their products for other than hand or horsepower. The gas engine required their cooperation to give it command of a new field.

Before the advent of gasoline engines, corn shelling and most other farmstead chores were performed by hand power. One exception was pumping water, which often was accomplished by harnessing the wind. Windmills were a common sight on American farms from the 1870's through the 1930's.

A view in the paint shop of the Olds Gasoline Engine Works in 1901.

54

The advertisements appearing here were taken from the Gas Review, an engine trade journal first published in 1908. It had a large circulation in its day.

In its first application to farming, the gasoline engine was used almost entirely to pump water for livestock, often replacing the windmill already in operation. Subsequently, it was called upon to pump water into the house for domestic use. There it became the women folks' best helper, turning the washing machine, cream separator, churn, and a dynamo for lighting the home and premises. Soon it was sawing firewood, grinding feed, and performing a dozen other chores around the farmstead. Thus, step by step, the range of usefulness and influence of the gasoline engine broadened.

The acquisition of that first gas engine was an occasion

With a gasoline-powered machine, the farmer's wife had leisure time on washday. (USDA)

to be long remembered, as one farmer related in the **Gas Review** in 1909:

... so taking all points into consideration I ordered a new engine. When it came and I got it home, all I had to do was hitch on the battery, put in some gasoline, fill the reservoir, or jacket, with water, turn the wheel and away it went. The whole family and half of the neighborhood were in attendance. I realized that there would be many sacrifices before it was paid for, but the feeling of having an engine of my own balanced up any thought of sacrifice I might be called upon to make.

Gasoline power was passing its experimental stage. New farm uses were increasing the demand for engines. In 1911 it was estimated that more than 100,000 engines were purchased strictly for farm use, ranging in size from 1½ to 20 horsepower. Farmers, who until a few months before had never seen a gasoline engine, were buying them. Others who had sworn they were fakes were feeling the need for them. The machinery sections at state fairs, with their popping engines, reminded one of small arms practice at the annual encampment of the National Guard. If making noise was advertising, gas engines were certainly doing the job right. Yet, with all this activity, farmers had much to learn about power machinery.

Prior to World War I, the principles of gasoline engine operation were frequently expounded in various trade journals. No machine since the self-binder had called for so much expertise in overcoming petty troubles. For years after its introduction, the binder was a considerable mystery to the majority of farmers, and during harvest time the dealers were besieged with questions. So it was now with the gas engine. More sophisticated than farmers were accustomed to handling, it required more than just approximate adjustments to remain operating.

A query of gas engine manufacturers disclosed that their most vexing problem was "the engine operator." In nine out of 10 cases, trouble could be traced directly to negligence or a lack of ability on the part of the operator rather than a defective engine. Many men would not exercise reason in locating a problem; instead they would produce a wrench or a pair of pliers and begin — somewhere. The results merely compounded the problem.

There probably wasn't a country crossroads anywhere that didn't boast of at least one, and usually several, "experts" — men who were not slow to confess with pride that they knew all there was to know about gasoline engines. In the end, many of these men succumbed to the old engine proverb: "An hour's search and three minutes' fixing."

Notwithstanding a vigorous campaign to promote engine sales, not more than 20 percent of American farms were equipped with them in 1913. Many farmers stuck to their old ways through ignorance rather than by choice. Of course they had heard of gas engines and had watched automobiles rushing past their farms, but they understood little about them and professed less interest. The average farmer, possessing limited mechanical knowledge, looked upon a gasoline engine much as he did upon a steam engine. It was a mighty fine thing but too complicated for him. Upon viewing a gas engine at work, he very likely shook his head and remarked: "Yep, it's wonderful, but I wouldn't know what to do with it. I'm no engineer!".

A few farmers regarded the gasoline engine as a stubborn curiosity, oftentimes possessed of the devil, running when it felt like it; not running when it didn't feel like it. The operator was thought to be helpless until the engine decided to go. To the uninitiated, an engine that refused to run was the most wrath-provoking, satanic contrivance ever thrust upon long-suffering man.

It was evident there was a need for an intelligent campaign to show the farmer that operating a gas engine was no more difficult than running a mowing machine. This was frequently accomplished through the farm's young people. The average boy 14 or 15 usually was interested in machinery. An engine appealed to his imagination. With a little instruction in fundamentals, he often became an enthusiast. Many times a farm was completely changed through a boy's influence. A farmer might refuse to adopt engines, claiming he was too old for anything so "new-fangled," yet he permitted his boy to have a small one.

Gas engines helped to keep many boys on the farm. But some, as soon as they were old enough, drifted away to the cities to work in factories or wherever they could obtain employment. The reason for this restlessness was often attributed to the father, who, being content to execute the farm work in the same manner as his father had, now expected his son to fall in line. When the boys were young, turning the grindstone or fanning mill by hand had always appeared to be a lot of fun, but once they were old enough to do it — all they wanted to, and more — it was no fun at all.

The farmer who decided to purchase a gasoline engine had his choice of either vertical or horizontal design and air or water cooling. Although vertical engines were lighter and required less floor space, they received less attention in later years. The trend slipped to the horizontal type, which had a more efficient means of cylinder lubrication, thus longer life. Air cooling eliminated the anxieties of freezing in cold weather, but it was considered practical only for small units.

Water cooling generally was accomplished in one of two ways — the thermo-siphon system or the open-jacket system. With the thermo-siphon system, heated water from the cylinder jacket rose and passed into a sizable steel tank where it was circulated and cooled. With the open-jacket system, a hopper, usually cast with the cylinder, held a small quantity of water which absorbed the engine's heat. Although hopper-cooled engines required more attention, owing to the frequent replacement of evaporated water, they were simpler and more compact than tank-cooled outfits. In later years, the open-jacket system was used almost entirely.

When the cool, crisp nights of autumn arrived, it was time to drain the water from the engine hopper after each day's work. The inexperienced operator often procrastinated, believing that "if it did freeze just a little, it certainly couldn't break that big iron casting." Frequently this tardiness cost him a cracked jacket, and perhaps a new cylinder head.

The most common form of ignition on gasoline engines was the low tension or make-and-break. It required a source of direct current, a coil, and a device called the igniter for breaking the circuit within the chamber at the proper time. A set of dry cell storage batteries supplied the current. Later, batteries were replaced by the magneto.

Those persons buying small vertical gasoline engines often preferred models which were air cooled. The need for water and a cooling tank was thus eliminated. This example was manufactured by the International Harvester Company in 1908.

Most gas engines were furnished with a mixing valve for carburetion. This device was much simpler than a carburetor. There were fewer parts and fewer adjustments to make. The "mixer" was eminently satisfactory on engines which ran at a constant speed.

Most engines used for farm work employed the hit-and-miss system of governing. This system was distinguished by the absence of one or more cylinder explosions when the engine's speed exceeded a certain point. A governor held the speed of the engine within defined limits by allowing successive charges or impulses only when the engine load required them. For this reason, the exhaust reports from a gas engine were irregular when under a light load, and became regular only when handling a full load.

It was common for the unsophisticated operator to either underestimate or overestimate the capacity of his engine. A safe rule to follow was "never run your engine with a steady load beyond two-thirds of its rated horsepower." Being rated at near their maximum load, gasoline engines had no reserve power. A farmer needed to calculate the amount of power necessary to operate his heaviest machine, then buy an engine of one-third greater capacity.

Engines frequently were mounted on hand or horse-drawn trucks to increase their mobility. It usually was more convenient, for instance, to take the sawing outfit to the woods than to haul the logs to the barn. An engine could be backed up to the kitchen porch to run the clothes washer in the morning, then wheeled over to the

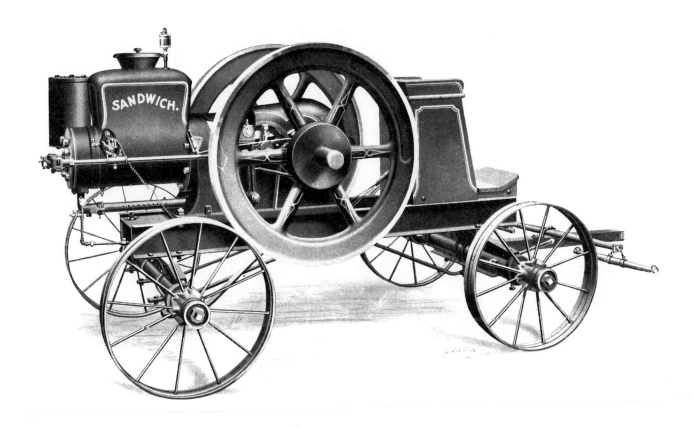

Portable gasoline engines were built in two different styles: hand-portable and horse-portable. Any outfit requiring the use of a team to transport it, such as this open-jacket, horizontal Sandwich, was called a horse-portable.

barn to operate the feed mill or corn sheller in the afternoon. Portable engines seldom ran out of work.

A few ingenious farmers built their own tractors by mounting a gas engine on a frame and improvising with an assortment of old gears and miscellaneous parts. The bull wheels from a couple of discarded binders could be requisitioned for drivers, and either mowing machine or binder truck wheels were used in front. Factory supplied "traction gearing" was sometimes used to simplify the conversion process.

Farmers often found it not only convenient but also profitable to erect a building specifically designed to meet their power needs. With a farm power house, several machines could be operated at once from the same engine, thereby saving time and fuel. Such a building also saved labor, since the engine did not have to be moved and reset for each different job. A common arrangement in the power house placed the engine, along with a few shop tools, such as grindstone and emery wheel, in a separate room in one end of the building. If a dynamo was used for home lighting, it also was located there. In another room, a feed grinder, corn sheller and fanning mill were situated near windows through which a wagon could be unloaded. Because of the dust created by such machines, this room had to be tightly partitioned. A third room, frequented by the housewife, contained such items as a washing machine, cream separator and churn. These machines occupied a fixed position in the building and were driven by belts from line shafting which was set overhead or hung from the ceiling.

This 1½-horsepower Fairbanks-Morse outfit was a hand-portable. It could be drawn from one job to another with little effort. It was just the right size to operate the grindstone, washing machine and other similar equipment.

Engines mounted on skids were known as semi-portable. They could be moved from place to place, but not as handily as those mounted on wheels. Shown is an Olds skidded engine of about 1909. Note the large thermo-siphon cooling tank.

Engines were sometimes placed on long steel channels which permitted them to be mounted on farm wagons. The Domestic Engine and Pump Company offered such an option for its engines of six horsepower or more.

This arrangement proved quite practical for use in farm powerhouses. With an engine of adequate size, several machines could be operated simultaneously. The gasoline tank usually was located outside and underground.

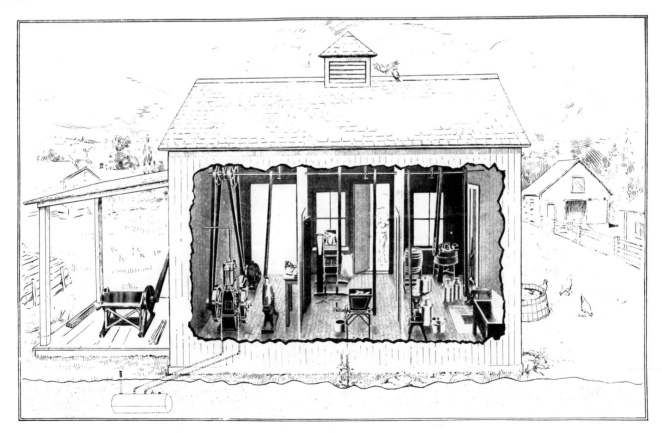

The style of gas engine selected was determined by the location and the nature of the work to be performed. Operating a farm powerhouse, for instance, required the use of a stationary engine. This example, manufactured by the Stover Engine Company, was available in sizes ranging from 4- to 12-horsepower.

Portable wood-sawing outfits were popular on the farm. The choice of either a four-, six- or eight-horsepower engine was available on this outfit, built by the Sandwich Manufacturing Company. Weight, with six-horsepower engine, was 2,510 pounds.

"Buzzing wood" in the 30's with a tilting saw table and McCormick-Deering gas engine. (International Harvester Co.)

The drainage of manpower from ordinary pursuits during World War I greatly hastened the adoption of labor-saving machines. Manufacturers turned out 600,000 new one-cylinder engines in 1917, topping any previous year's output by nearly twofold. There were now an estimated 2,000,000 units in use on American farms. Owing to the high price of gasoline, engines designed to burn kerosene were sold in increasing numbers, especially in the larger sizes. The "oil engine" had only recently been perfected after years of experimental work.

The general adoption of the hit-and-miss gasoline engine resulted in greatly improved methods of performing various farm tasks. The winter's woodpile was always a rapidly-decreasing commodity, and the labor expended in cutting it by hand was a job to be dreaded from year to year. Anyone who had bent his back over a bucksaw for a day knew the magnitude of preparing a season's supply. According to the United States Department of Agriculture, 86,000,000 cords of wood, valued at $250,000,000, were being consumed annually by 1911. Nearly 75 percent of this fuel was used on the farms where it was produced. With an engine of from three to six horsepower and a tilting saw table, a farmer could harvest 20 or 30 cords per day.

Many farmers used windmills to supply water to their livestock. A windmill was an economical and untiring source of power that saved countless hours of hand-pumping — a task that often fell to boys or the hired men. Mounted on a tower at least 15 feet above all buildings and other obstructions, this device could normally pump several hundred gallons of water per hour. But there were times when the air was so quiet that the wheel could not operate; and there were other times when high winds made its operation unsafe.

Because of his wife's fears that "the noise they make would break up the setting hens," a midwestern farmer purchased a windmill rather than the gas engine he desired. That windmill ownership could occasionally have its drawbacks is evidenced by his report in the **Gas Review**:

> She figured that when the windmill was installed the cost of upkeep would be practically nothing, and in a sense she was right, but as time passed many things turned up that we had not figured on. One day I went to the village to do some shopping and left the boys to grind the feed for the work horses. While I was at the village a cloud came up which brought with it a strong wind. I hurried home as fast as possible and when I came in sight of the mill it was running like a streak. I wondered why they had not shut it out of gear, but when I came to the barn I found that the triprod had worn in two and it was impossible to

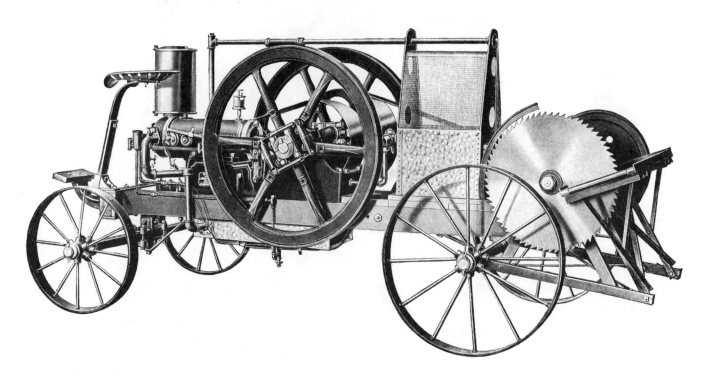

This wood-sawing outfit was manufactured by International Harvester. It featured a six-horsepower horizontal engine and a galvanized steel cooling tank. Sawing capacity was claimed to be four cords per hour.

This was a popular method of operating an ordinary well pump. A belt drive eliminated the need for extra gearing on the engine, allowing it to be removed at any time for use on other jobs. This engine is a two-horsepower Domestic.

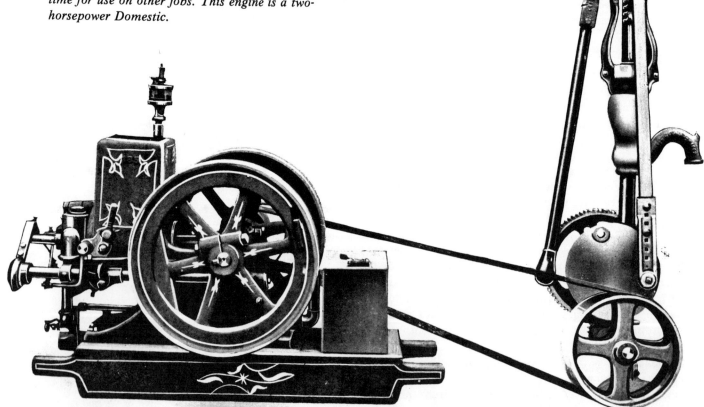

This IHC "walking beam" pumping outfit was a combined pump jack and engine, mounted on the same base. By means of skids, it was transported easily from place to place. The tank-cooled vertical engine was rated at two horsepower.

shut it out of gear without climbing to the top of the tower in the gale, which of course would be a dangerous undertaking. While we stood there deciding what should be done, fate solved the problem. The old mill made a sharp turn one way, and then the other, and out flew a section of the fan, and to cap it all, the section flew into the top of the new buggy that I had just unhitched from. One by one the fans were whipped out until nothing was left but the arms and vane of the mill.

In later years, a self-regulating feature was added to windmills that virtually eliminated the possibility of storm damage. If the wind reached a dangerous velocity, usually about 25 miles an hour, a governing mechanism came into play which automatically threw the wheel out of gear.

Although windmills possessed many advantages, pumping dependability was improved considerably when hit-and-miss power was extended to the farm well. The gas engine often worked in conjunction with the windmill, performing the pumping chores when the wind could not.

Prior to the 1920's, the farmer generally was still dependent upon kerosene lamps and lanterns for lighting his home and premises. Central-station electricity was confined to cities and nearby towns, having not advanced into rural sections because of the high cost of building lines and transmitting power. Coupled with the gasoline engine, the electrification of the isolated farm home became possible. Small plants were designed exclusively to place electricity at the disposal of the farmer. A contrivance of soft iron, copper and wire, called a dynamo (generator), driven by a gasoline engine, charged storage batteries during the day. At night the batteries were ready for lighting the house or barn.

Until the advent of the tungsten lamp, the small engine-driven electric plant was impractical because of the high cost of operation. The discovery of this economic filament, requiring only one-third of the current needed by the old carbon lamp, hastened the adoption of electricity on the farm.

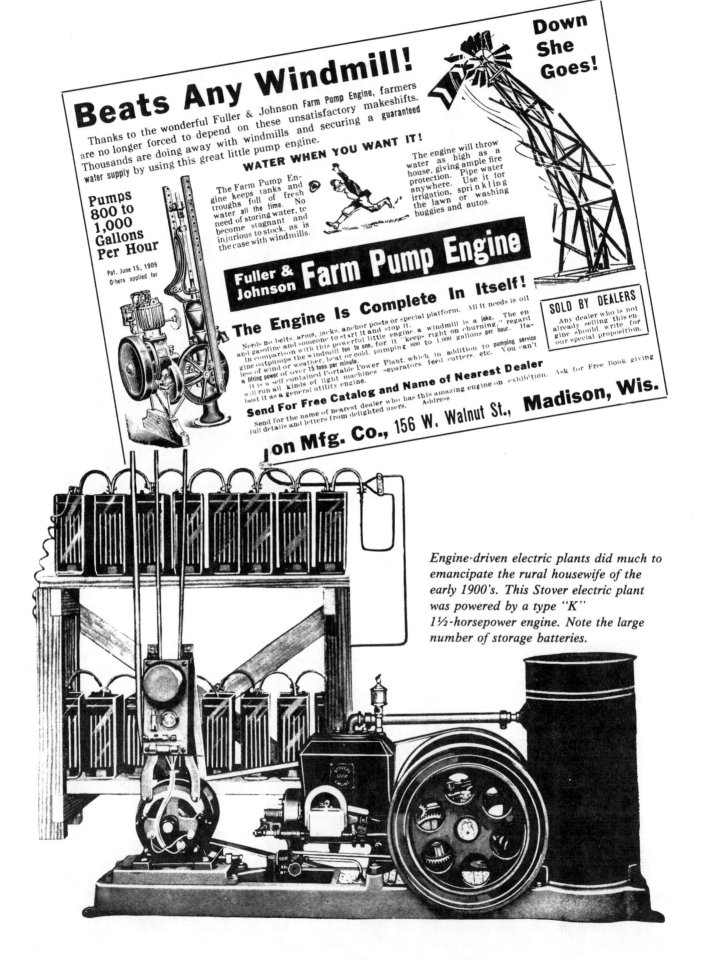

Engine-driven electric plants did much to emancipate the rural housewife of the early 1900's. This Stover electric plant was powered by a type "K" 1½-horsepower engine. Note the large number of storage batteries.

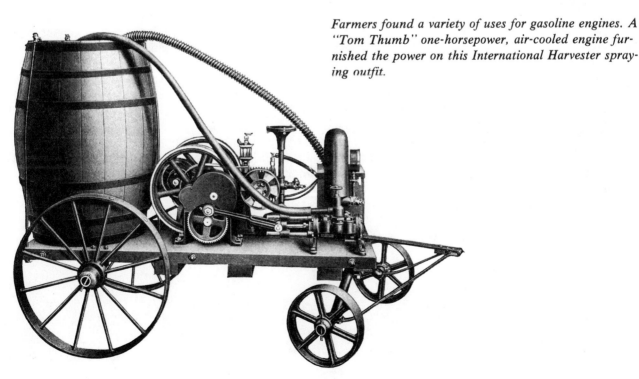

Farmers found a variety of uses for gasoline engines. A "Tom Thumb" one-horsepower, air-cooled engine furnished the power on this International Harvester spraying outfit.

A three-horsepower John Deere gas engine shells corn in 1930. (Deere & Co.)

The adoption of the gas engine for use on grain binders was an example of its versatility. During wet harvest seasons, it often became nearly impossible to drag binders through the soft fields. The 1915 season was one of the wettest on record, and thousands of acres that otherwise would have been lost were harvested in this manner. The cardinal requisite of a successful binder engine was light weight. One manufacturer produced a model weighing less than 170 pounds, while yielding four horsepower. The engine was attached directly to the main frame behind the bull wheel and, by means of a chain drive, lightened the horse's task to merely pulling the binder. Two horses could handle an eight-foot machine in heavy grain.

Farmers enjoyed the benefits of the gasoline engine off the farm, too. Blacksmiths often used gas engines to run band saws, power drills, forges and emery wheels. Sometimes they installed grist mills in their shops so that a farmer could get his corn ground into feed or table meal while he was having his horse shod or his plow points sharpened.

Although the one-cylinder gasoline engine was used considerably in the 1930's, it was rapidly losing ground to engines of advanced design. Of even greater consequence were the inroads being made by rural electrification.

By about 1925, central-station electricity was beginning to be a practical and economical source of power for an increasing variety of farm uses. More than 200,000 farms were receiving central-station electricity that year. By 1935 this number had increased to nearly 2,000,000, and electric lines were being extended to about 300,000 additional farms annually. A transition was in full swing. The hit-and-miss gasoline engine was in the final phase of its transitory existence.

4
Harnessing the "Modern Farm Horse"

In 1897 two young men from Minneapolis, O. B. Kinnard and A. Haines, explored the possibilities of building a gasoline-powered traction engine. Having their own machine shop and several years of mechanical experience, they succeeded in developing a one-cylinder machine and placed it on the market the following year. The Flour City was one of the first gasoline tractors in the country.

In 1899 one of these machines was exhibited at the Minneapolis fairgrounds, where its unusually heavy backfiring caused the crowds to retreat in disorder, fearing that it was "going to bust" like steam boilers did on rare occasions.

This original style one-cylinder Flour City tractor was manufactured in 1899.

Above: Illustration of Hart-Parr factory and first tractor.

Two young engineering students, C. W. Hart and C. H. Parr, of Charles City, Iowa, built a two-cylinder traction engine in 1902. The following year 15 oil-cooled units were sold. Their first machine was still in service 17 years later. Credited with producing the first really successful engine propelled by internal-combustion, the Hart-Parr Company was also the first to use the word "tractor," which, in 1906, began to replace the more cumbersome expression "gasoline traction engine."

The 17-30 Hart-Parr, built in 1903, was still operating in Minnesota 20 years later, when this photo was taken.

International Harvester produced this gasoline tractor in 1908. It was available in 10-, 12-, 15- and 20-horsepower sizes. Speed was regulated by a hit-and-miss governor. Water was cooled with a tank and cooling tower.

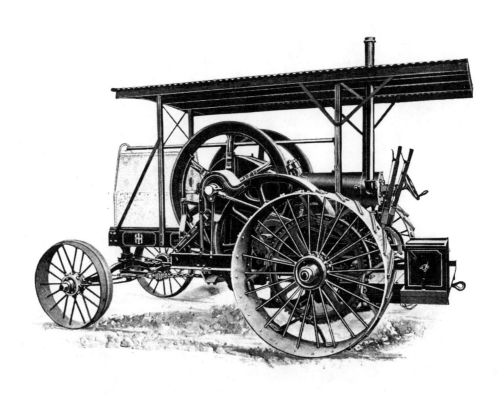

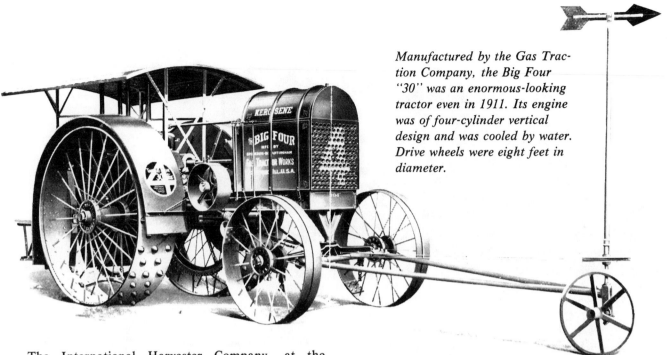

Manufactured by the Gas Traction Company, the Big Four "30" was an enormous-looking tractor even in 1911. Its engine was of four-cylinder vertical design and was cooled by water. Drive wheels were eight feet in diameter.

The International Harvester Company, at the forefront in the production of harvesting machinery, introduced a 20-horsepower single-cylinder tractor in 1906. Several larger sizes soon followed, establishing the company as a formidable power in yet another field.

Only a daring prophet could have correctly predicted the future in 1907. At that time, a mere 600 tractors were scattered throughout the country. The possibilities raised by this new form of power were scarcely being realized. The "horseless farm" sounded mythical and beyond the reach of any but those who pursued farming as a fad.

The increasing interest in tractor farming manifested by 1909 began to portend the role it soon would play in American agriculture. Admittedly it was a new idea, and like the transition from the ox to the horse, it was met by argument and counter-argument. However, the sale of 2,000 new units that year left little doubt that the internal-combustion tractor was emerging from its experimental stages.

The tractor placed in the farmer's hands the means of directly reducing production costs. Horses were not only expensive to buy but also costly to maintain. Each growing season about five acres of productive land had to be allotted to every horse in the barn, for raising oats, fodder, hay and straw. Sustaining the 24,000,000 horses and mules on American farms required a total acreage equal to the combined areas of Iowa, Illinois, Indiana and Ohio. It took a better-than-average horse to equal the value of his feed eaten over the course of one winter. In contrast, when the tractor didn't work, it didn't have to be fed.

The farmer's work was usually a race against time. In many areas, fully one-third of the "possible" plowing time was lost each year because of climatic conditions. Each farmer had to maintain extra work animals to meet emergencies brought on by inclement weather. But getting through the rush season entailed the services of more than just supplementary teams; additional men also were necessary.

The horse's weakest point was its endurance. Ten hours a day was often considered a horse's working period. However, authorities claimed that eight hours were better, and that six hours under a heavier load would accomplish the same volume of work with less physical wear and tear. A typical animal could not be relied upon for an average of more than 13 to 15 miles of pull a day or more than six hours of work a day, even during the busiest months.

Spring work had to be rushed at top speed. Horses at that time of year were soft, and many suffered from panting flanks and galled shoulders. Tractor endurance was an unknown quantity. At the close of the day it was as fresh as it was at the beginning, and there was no lagging in the traces during the late hours of the afternoon. In fact, it could be operated day and night, if need be.

Equally important was the fact that when the day's work was finished there were no chores to be done. Horses had to be unhitched, driven to the barn, and their sweaty harness removed. They also had to be watered, fed, curried and bedded down. Care of a horse was said to require 27 minutes of a man's time each day. With a tractor, when the rattle of the exhaust ceased and the flywheel slowed to a stop, work was done for the day. And in the morning, instead of going to the barn, feeding the horses and cleaning their stalls, the farmer could proceed to the field where his "modern farm horse" was standing, give its crank a turn and start the day's work. He had to fill the fuel tank and turn some grease cups occasionally, but compared to the steady day-in and day-out routine of caring for horses, there was nothing to it.

At the Winnipeg Motor Contest held in August, 1911, Flour City tractors won gold medals in two of the three trials in which they were entered. This 40-horsepower tractor won top honors in the kerosene class.

By 1909 the gasoline tractor was causing concern in the ranks of the old-line steam thresher companies. After years of mediocre success, the new and struggling tractor firms were now reaping some profits on their investments. Watching this development with increasing alarm, steam men concluded that it was time to act. At first they had observed the crude gas tractor on exhibition at fairs, cracking and snapping like a gattling gun, backfiring and bucking like an unbroken bronco and they had felt pity for its unfortunate builders. Now it was different. Although some were reluctant to admit it, the tractor was cutting a considerable swath as a farm motive power. Many steam manufacturers were quietly working on experimental gas machines which would soon be ready for market. They had remained out as long as they dared and were now getting in line for self-protection.

The saving of time and labor was the big factor. The gas tractor did away with the necessity of a water tank, a water boy and an extra team; with dirty and leaky flues, burned out grates, bad water and dirty boilers. The engineer could safely sleep as long as the rest of the crew instead of rising a couple of hours earlier to get up steam. With the turn of a crank the tractor became a thing of life, capable of performing its task at any hour of the day. The engineer did not have to bank the fire, clean the flues, or blow off steam to clean the boiler for another day's run.

Then, too, the gas tractor was not as heavy as the steam engine of equal horsepower. This was an advantage in plowing, as objections often were raised against the way heavy machines packed the ground.

Steam power often was hampered by difficulties that could be overcome by the gas tractor. In many cases where steam plowing was desirable, coal was either not available or was too expensive. Or, perhaps the water was unfit for boiler use or had to hauled too far. One filling of a tractor's fuel tank was sufficient for a day's run. Then, too, fires were practically eliminated by gas power, no matter how dry or windy the conditions might be.

Notwithstanding the advantages of gas power, it was not totally successful . . . at least not yet. Many tractors were somewhat unreliable, and those that did work well were considered far from perfect. Upon delivery of a new tractor, a factory expert customarily was present to make whatever mechanical adjustments were necessary. Often there were problems which required him to remain for days. A few tractors were so difficult to start that operators let them run all night rather than face almost certain exasperation in the morning. Experimental tractor design resulted in frequent breakdowns and large repair bills. However, the development of the industry was remarkably rapid, with one improvement after another resulting in greater reliability.

By 1911 the future of the gas tractor in agriculture was no longer in doubt. With the supply of new tractors unable to meet the demand, manufacturers were doing a thriving business. Many expanded their equipment and shop facilities. The M. Rumely Company completed a new building measuring 650 feet in length for manufacturing the "OilPull." The Hart-Parr Company increased its production capacity to 2,500 tractors annually. That the industry was becoming firmly established was demonstrated by the sale of more than 7,000 tractors, worth more than $16,000,000, by year's end.

Most tractors of the period were leviathan in size — copied generally after their steam predecessors. Large drive wheels, ranging from six to eight feet in diameter, were popular. Fuel tanks holding 60 to 110 gallons were the rule. Even with lower weight per horsepower compared to steam engines, the machines were still giants. Designers believed that there could be neither power nor traction without great weight.

New design engineers favored four-cylinder engines. However, the slow-speed one and two-cylinder machines remained popular. These engines required a large flywheel to store up energy received by the crankshaft and to deliver it in a uniform manner.

While many earlier tractors had flyball governors of the hit-and-miss type, along with make-and-break ignition systems, the throttling governor and jump spark method provided a steadier motion and was coming into general use. Electric current for starting was supplied by dry cell batteries. Once the engine was running, a low-voltage magneto took over.

Some tractors were provided with closed radiators of the automobile type. Others employed the tower cooling principle in which water was pumped over a screen and allowed to drip into a tank or receptacle below. Oil-cooling systems, with obvious advantages in both the prevention of freezing and the overheating from evaporation, also were popular.

The lack of a uniform standard for rating horsepower was often a disturbing element in the early 1900's. Steam manufacturers had formed the habit of giving "nominal" ratings, which represented only about one-third to one-

Above: The M. Rumely Company introduced the OilPull tractor in 1909. Three years later, these machines were being manufactured at the rate of 50 a week. The 30-60 OilPull in this 1912 scene is drawing an eight-bottom engine gang plow.

Below: The Twin City "40" was an excellent "puller." In this scene it is handling twelve 14-inch plows near New Rockford, North Dakota.

half the actual horsepower their engines could develop. There was also an absence of uniformity in these ratings; the 20-horsepower engine of one company was often as powerful as the 22-horsepower engine of another.

This practice of under-rating dated back to the days when steam power was first used for driving threshing machines. Then an engine was said to have 10 horsepower if it could produce energy equal to 10 work animals on a horse-driven rig. Nominal horsepower, therefore, gave only an approximate idea of the true size of an engine.

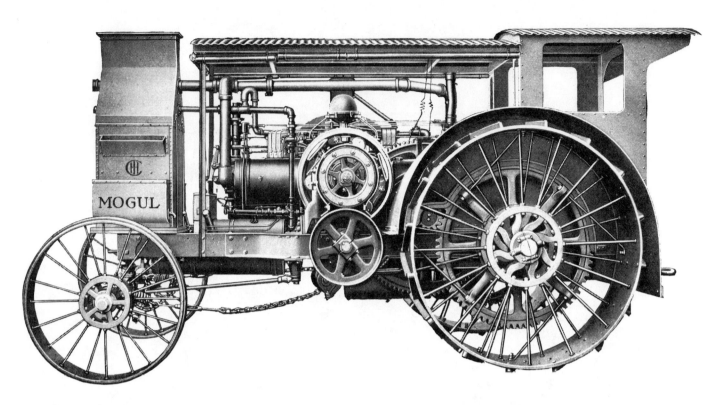

Above: The difficulty usually encountered in starting a large tractor in cold weather was overcome on this IHC Mogul, two-cylinder opposed 30-60. A one-horsepower starting engine relieved the operator of turning over the flywheel by hand.

Below: The Hart-Parr "Old Reliable" 30-60 tractor was manufactured from 1907 to 1918. A 300 r.p.m. two-cylinder horizontal engine furnished its power. One 2.3 m.p.h. forward speed and one reverse speed were controlled by a single lever.

The 40-80 was the largest gas tractor in the Avery line. Its engine was of four-cylinder opposed design and was cooled by oil. Weight of tractor was 20,000 pounds. Price in 1914 was $2,650.

Conversely, internal-combustion engines were always rated at near their maximum load, as determined by a brake test. This left many tractor men in a quandary since they could not understand why their tractor engine didn't develop as much power as their neighbor's steam engine of the same rating. Although the confusion was reduced in time, it was never entirely eliminated during the steam era. A few steam manufacturers preferred to indicate the size of an engine by cylinder dimensions; for example, a "9 by 10" rather than a "15 horse" machine. The practice of giving two ratings was applied to the gas tractor. One corresponded to the drawbar horsepower and the other to the belt. This was expressed, for example, as 22-45 or 30-60.

The tournaments held in the days of chivalry when knights in armor met in mortal combat were never more hotly contested than the annual motor trials inaugurated at Winnipeg, Canada, in 1908. The most proficient engineers, operating dozens of engines with the latest improvements, vied for several days in a struggle for supremacy. It was the ultimate test of man and metal, of skill in operation and cunning in construction.

The event drew thousands from all parts of the world to weigh the problems of tractor design. Representatives from all of the engine manufacturers were there, observing, studying and commenting. Farmers were also present — intently watching the activity.

Both steam and gasoline tractors were entered. They were judged by a point system based on their respective performance in two general contests. Each engine was put on a brake test to determine its horsepower as well as its economy in stationary belt work. Maximum horsepower was actually not sought, but rather the greatest horsepower consistent with the most economical consumption of water or fuel.

The most spectacular part of the trials took place in the plowing field. This consisted of a level stretch of the toughest Manitoba gumbo, a black, waxy soil that could thoroughly test the mettle of any engine. In addition, judges examined the relative merits of the design of each machine. Until discontinued after the 1913 trials, the Winnipeg Motor Contests served as a public demonstration of rare value to the farmer, designer and manufacturer.

By 1912 the tractor industry had prospered for about six years. Nearly 12,000 machines were built that year. One firm alone sold enough tractors to displace 60,000 horses. By 1913 the market was oversupplied. Hundreds of farmers had purchased tractors who had no possible means of making them pay. Finding it difficult to cope with mechanical problems which frequently arose, many famers were unable to keep their machines in running order.

Experiments with heavy tractors on old land in the corn belt were not entirely satisfactory. The wheels caused serious damage to the soil, and, of even greater consequence, farmers could not afford the machines unless they did a lot of custom work each year in addition to their own. Furthermore, farmers often did not know how to adjust their operations to take full advantage of the new power. As a result, the tractor market slumped over the next two years.

The two-cylinder Waterloo Boy, introduced in 1912, was one of the first all-purpose tractors on the market. It was a pronounced success from the beginning and went on to be manufactured for several years. The Waterloo Gasoline Engine Company was absorbed by John Deere in 1918. (Deere & Co.)

The big tractors generating from 30 to 40 tractive horsepower had been acccpted enthusiastically by the prairie pioneer — the man who went west of the 100th meridian or into western Canada. These men needed all of the power they could get to turn over the miles of tough buffalo grass. Considering the vast acreage to be plowed and the futility of attempting it with horses, manufacturers were given a wide latitude in the matter of price. The western farmer wanted service and was willing to pay for it. For this reason, tractor manufacturers devoted most of their energy to developing the heavy outfit.

But as the new territories were settled and the land became broken, the demand for large tractors declined. A new and greater demand emerged for a medium or light-weight tractor, not only for western farms, but also for lands in the corn belt and the East. Unless a tractor could be used on a variety of jobs, like the horse, it could not hope to benefit the average farmer. By the summer of 1914 the main topic of conversation in power farming circles was the development of the small general-purpose tractor.

Expenditures for power on the all-horse farm were increasing in 1914. There were 25,000,000 work animals in the United States, and their estimated average value was

Allis-Chalmers' first entry into the gas tractor field was this two-cylinder 10-18, introduced in 1914. Here it takes part in a plowing demonstration in North Dakota. (Allis-Chalmers Co.)

$109. That represented an increase of 143 percent in the value of the average horse since 1900. The figure, however, included colts and range horses. For a really good draft animal, a farmer usually had to pay from $140 to $200.

Farmers were demanding a tractor that could take the place of five horses — one that could pull a two-bottom plow and work untiringly during July and August when a team could not stand the heat and pestering flies. Most farmers spent as much time plowing each year as they did in all other types of field work combined.

Riding up and down the furrows on the iron seat of a sulky plow all day long, keeping three horses up to the grind, and then turning over only two or three acres, was mighty monotonous. The average farmer had only about 80 acres of small grains, and a two-plow tractor would enable him to turn his ground at the proper season.

Deep-plowing was advocated by many agricultural authorities. A canvass of state agronomists indicated that, except in a few favored areas, the depth of plowing should be doubled. Although deep-plowing with horses was feasible, its greater demands upon animal power made it impractical. When the days were long and hot and the ground was hard and dry, it was especially difficult to maintain a satisfactory plowing depth with

The Oliver No. 62
Engine Gang Plow

Turning the soil with an IHC Mogul 8-16 and Oliver gang plow. This tractor, introduced in 1914, was powered by a single-cylinder horizontal engine which used make-and-break ignition.

In its advertising literature of 1916, the Huber Manufacturing Company appealed to the farmer to "buy a tractor big enough." This four-cylinder 35-70, with its 96-inch drive wheels, was nearly as big as they came.

horses. The sulky or gang lever often had to be raised a notch or two during the heat of the afternoon. Sometimes the work had to be delayed to wait for a rain — or for the shoulders of the horses to heal. Mechanical power in the form of a tractor would permit the farmer to plow as deep as he wished and when he wished.

The extent of the new demand for small "one-man" tractors was illustrated graphically at the Fremont, Nebraska, plowing demonstration in 1914. Trains and automobiles brought in an estimated 25,000 farmers, the majority of whom expected to eventually buy a tractor. Fully half of the more than 50 tractors exhibited were the small all-purpose type. The big, cumbersome machines were giving way to the two- and three-plow outfit that was easy to handle and didn't cost as much as the farm it was intended to cultivate. Farmers were saying that what the "old reliable" had accomplished for the big power users, the small outfit would do for them on a restricted scale.

Although farmers generally were acquiring a mature attitude about the advantages of power farming, the idea did not take the country by storm. To many, the idea of discarding the horse was considered unthinkable and impossible. One farmer presented his views in the **Gas Review**:

Since reading statements regarding plans to relieve Old Dobbin of his usefulness on the farm by replacing him with tractor engines, I have been doing some thinking. I have a member of the Old Dobbin family that is of the third generation which, together with his two direct forbearers, have rendered constant and unrequited service in our family for practically 60 years — and the possibility of not continuing that co-partnership presents a problem I just don't want to contemplate. The horses on our farm are not only animals of servitude but companions and friends, partners in the business of solving life's problems, surrounded by a sentiment of love and mutual dependence that cannot be broken . . .

Despite their waning demand, large tractors of great power were not to become obsolete. They had been tried and found true to their advertised merits in a dozen years of use. For powering huge threshing machines, their utility would extend through the 1930's. Also, another branch of service for which they were well suited — highway improvement — was opening.

Above: The IHC Titan 10-20 tractor, introduced in 1914, was equipped with a horizontal two-cylinder valve-in-head engine. It was produced in considerable numbers until supplanted by the four-cylinder McCormick-Deering 10-20, in 1922.

Below: In 1916, the Moline Plow Company advertised its two-cylinder Moline-Universal tractor as being "as powerful as five horses; as enduring as seven horses; costs less than four horses; requires less care than one horse; less room than one horse; and eats only when it works."

The small all-purpose tractor was being sought with unabated enthusiasm by 1915. The potential buyer had a wide variety from which to choose. He could select a two-wheel, a three-wheel, or a four-wheel outfit. If he was particular about how it should be driven, he had just as wide a choice. There were tractors with a single drive wheel running in the furrow, or on the unplowed ground; also tractors with two, three, and four drive wheels. A farmer could have the drive wheel in front, or behind, or off to one side.

Numerous attempts were made to produce small tractors. Scarcely any two were alike — even in minor details — making any sort of classification extremely difficult. Although many machines offered salient features, there were undeniably a lot of misfits.

All this diversity indicated that both designers and manufacturers were at sea on the best way to propel a tractor. In the early days of automobile development there were many styles. One never knew, for example, whether the engine could be found under a hood in the front of the car, under the footboards in the center, or in some out-of-the-way position in the rear. The tractor was now where the auto had been. Once the problems of design were resolved, it would be merely a matter of refining the details.

The difficulties were centered around side-draft and traction. The lighter the tractor compared with the draft, the greater the need to have the center line of traction match the center line of draft. This is why so many tractors had three wheels with a single drive wheel. With this design, the side-draft could be eliminated by hitching directly behind the drive wheel. There was also the matter of stability. If the single drive wheel was high, there was danger of upsetting when turning a corner or on a side hill. If the driver was low, a serious loss of traction would result. The four-wheel tractor with two wheels acting as drivers was thought to have many advantages, but the side-draft inherent in this design tended to delay its general acceptance.

The cost of a new tractor in 1915 ranged from less than $400 to nearly $4,000. All the established companies that had been manufacturing heavy tractors were introducing one or more small machines in the eight to 20-horsepower class. At last the entire range of sizes had been covered. If a farmer didn't want a large tractor, he could get one weighing no more than a good team of horses.

Tractor engines also varied considerably in design. In 1916, 11 different types were being made, ranging from one-cylinder to six, in all styles of vertical and horizontal positions. There were vertical engines set crosswise of the frame, others set lengthwise. There were twin engines set both vertically and horizontally, single-cylinder vertical engines and single-cylinder horizontal. Then, there were both two-cylinder and four-cylinder opposed engines. In fact, virtually every known type of engine was being used

by some manufacturer. Which was the best-adapted for tractor use? Who could tell? Ask any sales manager and he could reel off 57 reasons for buying his particular line. If variety was the spice of life, the tractor certainly was highly seasoned.

As the role of the farm tractor expanded, the need for an improved distribution system became evident. Steam threshing machinery had always been sold by soliciting orders directly from factory or branch offices. When the small tractor came on the scene, manufacturers soon realized that direct selling methods were too expensive. The cost of selling a $500 gas machine was about as great as for a $3,000 steam rig. The question was who should distribute them? Although the automobile dealer was considered, it was the general implement dealer who was most familiar with the farmer's needs. He also usually had a mechanic who could be spared to leave the place of business for farm visits.

Before the internal-combustion era, the kerosene (coal oil) can was a familiar object in every household. Kerosene was used universally in lamps and for kindling fires in coal-burning stoves. Gasoline, on the other hand, was a by-product of kerosene and was almost a drug on the market. When the automobile leaped into prominence in the early 1900's, the demand for gasoline accelerated, as did its price. As a result, many tractor manufacturers turned their attention to the development of engines which could burn kerosene effectively. For engines to make efficient use of kerosene, it was found necessary to supply both heat and water to the fuel charge. Manifold heat was needed to provide thorough vaporization, and water injection served to keep the cylinders cool and free from carbon. With the price of gasoline nearly double that of kerosene during World War I, most manufacturers were turning out tractors designed for the cheaper fuel.

Forty-eight tractor companies exhibited their wares under the canvas tops at Fremont, Nebraska, in August, 1917. More than 300 tractors were parked behind a long double line of tents that made up the main street of the tent city. Total value of the exhibits was estimated at more than $1,500,000.

Never before had a demonstration drawn together such a galaxy of prominent tractor representatives. Many predicted the tractor was entering its final stage of development. The most striking thing, even to the casual observer, was a general similarity of lines in the new machines.

In 1918 tractors replaced 1,500,000 horses and 250,000 men who had gone "over there." More than 140 companies had machines on the market and sales exceeded 130,000 units, nearly double that of the previous year. The unparalleled demand for food and the shortage of available horsepower and manpower during the war prompted a substantial increase in the use of tractors on the farm.

Above: For several years after it was introduced, the low-priced Fordson represented 50 percent or more of the total tractor production in the United States. Its compact size made for easy handling on every farm job. The Fordson in this 1920 scene grinds corn near Dixon, Illinois.

Below: The International 8-16 tractor was first manufactured in 1918. It featured spark and throttle levers which were located on the steering column, like most automobiles of the period.

The Fordson tractor was introduced in 1918 after much experimentation. It featured unit-frame construction, an idea that was soon adopted by most manufacturers. The built-up frame, commonly used to support the working parts of early tractors, was eliminated. With the Fordson, various units including the engine, clutch and transmission could be removed with minimum time and labor.

The same year the Fordson was introduced, the International Harvester Company came out with the power take-off, which eliminated the need for a ground drive wheel. This feature permitted the direct transfer of power from tractor to drawn machine and paved the way for implements which had greater capacity and speed. It also did away with slipping bull wheels and with clogging and stalling where the ground-gripping action of the machine was not sufficient to run it.

The average tractor of 1920 incorporated many engineering advances that have remained standard since. These include enclosed transmissions, enclosed cooling systems, air cleaners, high tension magnetos, anti-friction bearings, pressure gun grease fittings, and force-feed engine lubrication.

In the early 1920's, few persons doubted that two tractors would be necessary for the efficient operation of a farm without animal power — a conventional model for general farm work, and a light model for cultivating row crops. A few companies were producing motor cultivators in either one- or two-row styles, but they were never well liked due to their high initial cost and limited range of use.

In 1924 the International Harvester Company introduced the Farmall, a tractor that revolutionized the industry. It featured high rear-wheel drive for maximum

This IHC Titan 10-20 tractor drives a hay press in 1918. All-purpose tractors usually were kept busy at some job on the farm year around.

axle clearance, narrow front wheels designed to run between rows, the means of mounting implements to front and rear, and the ability to turn within an eight-foot radius. The Farmall not only performed the belt and drawbar functions of the standard tractor but it also could cultivate row crops efficiently. It first made possible the horseless farm.

More than 500,000 tractors were on American farms in 1925. The four-wheel machine, with two serving as drivers, was dominating the market. Standardized design was spelling the doom of countless companies that had decided to "cash in" on the boom of the previous decade. The tractor industry was leaving behind it one of the largest graveyards of monumental failures of any industry in modern times — but out of it all came success.

During the depression years of the early 1930's, manufacturers made every effort to improve tractor design and boost their lagging sales. They spent money experimenting with wheel lugs of different shapes, sizes and spacings in an effort to increase efficiency, but with little success. Steel lugs provided traction, but they also damaged meadows, tore up barnyards, and were banned from improved roads. Furthermore, considerable power was required to thrust them into the ground and draw them out.

Large tractors of great power were by no means obsolete in 1919. This Aultman & Taylor 30-60 could operate the biggest grain separator on the market with power to spare. The engine was of four-cylincer vertical design. Extreme height to top of exhaust was 11 feet, 4½ inches. Diameter of drive wheels was 90 inches.

The Lauson 15-25 tractor of 1919 had modern lines for its day.

84

Above: The Minneapolis Steel and Machinery Company manufactured a popular line of tractors for many years. This Twin City 40-65 had a four-cylinder vertical engine, with each cylinder cast singly and mounted separately on the crankcase. The tractor's radiator held 130 gallons of water.

Below: This tractor, built by the Minneapolis Threshing Machine Company, was rated at 35-70 horsepower. The engine was of four-cylinder horizontal design. The radiator consisted of a top and bottom water tank, connected by a series of long brass tubes, cooled by a fan driven by belt from the motor crankshaft.

Above: OilPull tractors were advertised to be "as steady as a steamer." Like all tractors built by Advance-Rumely, this 20-40 OilPull had two cylinders and was cooled by oil. It could drive a 32-inch grain separator with all attachments.

Below: For years the largest and most powerful gas tractor manufactured in the United States was the Twin City 60-90. Although its design was similar to the Twin City 40-65, the 60-90 was more massive in order to transmit the enormous power of its six-cylinder engine. It weighed 28,000 pounds.

Above: The Allis-Chalmers 6-12 general-purpose tractor was powered by a Le Roi four-cylinder engine. It weighed 2,500 pounds and had a maximum speed of 2½ m.p.h. In this scene, it is pulling two 10-inch plow bottoms. (Allis-Chalmers Co.)

The Flour City 40-70 was a rugged machine built to handle the biggest power jobs. The motor was of four- cylinder vertical valve-in-head design.

Drive wheels were eight feet in diameter.

In 1932 the Allis-Chalmers Manufacturing Company introduced a feature that opened a new era in tractor mobility and efficiency — pneumatic tires. Mounted on rubber, a tractor could pull a 25 percent greater drawbar load, with a corresponding reduction in fuel consumption. The tires not only increased the life of the tractor and the comfort of the operator, but also made it possible to complete field operations up to 50 percent faster. The farmer now had a decided advantage in his annual race with time and weather.

In the years that followed, manufacturers turned to the production of the tricycle-type tractor with adjustable wheel tread, individually controlled foot brakes, hydraulic power lift, and the option of pneumatic tires. By 1937 nearly half the new tractors were on rubber.

With the basic machine constituting the modern tractor now complete, the stage was set for the mass exodus of Old Dobbin from the agrarian scene.

Above: Motor cultivators were on the market for several years but never really became a success. This six-cylinder, two-row machine was manufactured by the Avery Company.

Below: The Huber "Super Four" tractor of the early 1920's was rated at 18 horsepower at the drawbar and 36 at the belt. It was light enough to work plowed ground, yet heavy enough to drive a 28-inch thresher.

*The McCormick-Deering 10-20, introduced in 1922,
was for years one of the most popular tractors on the
market. Its one-piece main frame formed a substantial
foundation for the engine and a dust-proof and oil-
tight housing for all the working parts.*

*In 1923 the Hart-Parr Company built this compact,
all-purpose "20" tractor. It featured a two-cylinder
horizontal engine and water cooling. All Hart-Parr
tractors introduced after 1917 were cooled by water.*

Above: In 1924 the John Deere two-cylinder Model "D" tractor was put on the market. Farmers welcomed it because it gave them a combination of ample power, light weight and extreme simplicity. The Model D superseded the Waterloo Boy in the John Deere tractor line.

Below: The Farmall was the first successful attempt at building a genuine all-purpose tractor of the tricycle design. This tractor eventually was designated the F-20 in the IHC line. (International Harvester Co.)

Above: The "Light Weight" 30-60 OilPull of the late 1920's was one of the few really large tractors still being manufactured at this time. It was powered by a two-cylinder horizontal engine with a nine-inch bore and 11-inch stroke, which ran at a speed of 470 r.p.m. Note the huge flywheel and the hand starting lever.

Below: From 1918 to 1928, the J. I. Case Company manufactured a line of tractors with four-cylinder engines, mounted crosswise. The Case 18-32 shown here was one of the last tractors to follow this design.

Massey-Harris manufactured the Wallis "Certified" tractor for several years. Its patented U-frame contained in one piece the oil reservoir, crankcase, transmission case and supporting main frame.

J. I. Case discontinued its line of crossmotor-style tractors in 1929, when the Model "L" was introduced. This tractor could pull a three-bottom plow or operate a 28-inch thresher.

Above: In the early 1930's the Huber Manufacturing Company built several sizes of four-wheel tractors which were popular for threshing. This Model "HK" was rated at 52 horsepower on the belt.

Below: Pneumatic tires were replacing steel lugs in 1933. This Allis-Chalmers Model "U" tractor could pull three plows at five miles an hour and travel on roads at 15 miles an hour. Horsepower rating was 19 at the drawbar and 30 at the belt.

Above: The John Deere Model "A", introduced in 1934, was a typical tricycle-type tractor of the period. It could plow up to ten acres a day with a two-bottom plow, drill 25 acres a day with a large grain drill, or cultivate 30 acres a day with a two-row cultivator.

Below: Although most farmers were buying tricycle-type tractors in 1937, machines of four-wheel design still were popular. The Oliver Farm Equipment Company offered the "80," the "90" (pictured), and the "99" for big-acreage farming.

Old Dobbin was being relieved of many of his farm duties by the late 1930's. Cultivating corn was one of them. (Deere & Co.)

The STANDARD Model "E" TRACTOR
Steel Wheels With Spade Lugs

➤ Full four plow.

➤ Ideal for belt work. Will handle a 32-inch grain thresher and other belt driven machinery requiring equal power.

➤ Belt pulley on right-hand side. Easy to line up with belt driven machine. Ample belt clearance between front wheel and frame.

➤ Two speeds forward—2½ and 3¼ miles per hour—one reverse.

➤ Powerful 4-cylinder, heavy duty engine of Allis-Chalmers design; removable cylinder sleeves; full force feed lubrication; efficient water cooling system; unit construction; cut steel, hardened transmission gears; anti-friction bearings throughout; fully enclosed and protected from dust, with all important drives running in oil.

➤ A strong, rugged tractor, capable of handling toughest jobs, day and night. Powerful and steady on the belt. Can be equipped with air tires if desired.

➤ Horsepower—27, drawbar; 42, belt. Weight, 6,000 pounds.

Also available with 50 H. P. and 60 H. P. engines for belt work only.

Write for catalog giving full details.

AIR-TIRED MODEL "E" TRACTOR THRESHING

ALLIS-CHALMERS

—9—

5
From Leather Reins to Steering Wheels

The team and wagon were once a common sight on American country roads wherever farm products were transported to market. This wagon is a Weber, manufactured by the International Harvester Company.

H e who felt no affection for a farm wagon never spent his childhood in the country. Whether going to town for supplies, afield for threshed grain, or the woods for fuel, no greater thrill awaited a youngster than the unassisted climb up the wagon wheel to perch on the spring seat with Dad or the hired man. The jolt, the jar, the rumble of the vehicle in protesting the harsh roadbed, all added to the invigorating delights of rural living.

No other implement on the farm was so indispensable. Scarcely a move could be made during spring planting, throughout fall harvest, or in the conveyance of crops to market, without bringing the farm wagon into use. Occasionally it would groan if abused by overloading, or shriek when its axles were allowed to run dry, but when fairly treated it served its owner well, year-in and year-out.

The unparalleled demand for military vehicles during the Civil War spurred a rapid growth in wagon manufacturing. Formerly, when a farmer needed a new wagon he usually had one fashioned by the village blacksmith. In their formative years, each manufacturer produced a vehicle that conformed to local requirements. The requisites considered necessary in wagon construction differed from one part of the country to another, resulting in a diversity of types — each peculiar to a particular area. The vehicles carried in a Buffalo, Saint Paul, or Fort Worth repository, for example, were built according to specifications that were believed correct for the surrounding region. Wagons that met eastern requirements differed from those demanded in the Midwest or the South.

Wagon manufacturers prospered with an ever-expanding market during the last quarter of the 19th century. Many began incorporating additional types into their lines. Some of these lines eventually included wagons adapted to every section of the country. Commonly listed in manufacturers' catalogs were such types as Eastern States, Central States, Southern States, Cumberland-Blue Ridge, Southwestern States, and Rocky Mountain.

Above: This Capital wagon, built in Ionia, Michigan, was a typical marketing vehicle of the 1890's. Height of front and rear wheels was 44 and 54 inches, respectively. Tire width ranged from 1 3/8 inches on light vehicles to two inches on heavy ones. Price of the average weight wagon with double box and spring seat was $65.

Below: The Studebaker Corporation was the world's largest manufacturer of wagons and carriages. In 1910 there were more Studebaker farm wagons in use than any other three makes combined. This Central States wagon was one of the firm's best sellers. Triple-section box was available at extra charge.

Production and distribution of this wide variety of vehicles was no small undertaking. To add to the manufacturer's headaches, a prospective customer could specify that any detail be changed on his wagon or that supplemental features be added. Thus, an unlimited number of variations were produced, with each representing an increased cost to both the manufacturer and the consumer.

The 20th century brought many refinements in farm implement design — among them the standardized wagon. By producing three general types, manufacturers were able to reduce confusion, meet universal requirements, and offer a high grade vehicle at minimal cost. The three types were: Farm wagons for ordinary farm use, Mountain wagons for heavy duty in rough country, and Valley or Western wagons for intermediate grades of service between farm and mountain.

Manufacturers also began making an interchangeable line of parts and standardizing general specifications.

Wheels, once available in diameters measuring up to 54 inches, were usually standardized at either 40 and 44 or 44 and 48 inches (front wheels were always smaller than rear). Light wagons were sometimes equipped with wheels measuring 36 and 40 inches.

Wagon boxes were generally 10 feet, six inches in length and 42 inches in width.

The wagon's undercarriage, commonly called the gear, was now produced in four weights: light, medium, standard, and heavy. Capacities were usually 1,500, 3,000, 4,500 and 6,000 pounds, respectively.

The Mountain wagon was not merely a heavy farm wagon, but a vehicle designed especially to meet the exacting requirements of mountain road service. A heavy-

Farmers who lived in the South generally used Southern States wagons. These vehicles were sometimes equipped with boxes having special dimensions for hauling cotton. On this Studebaker, a spring seat was not included as part of the regular equipment.

duty brake was standard equipment, in addition to a double-braced gear and reinforced bed. These sturdy vehicles occasionally weighed more than a ton and had capacities up to 8,000 pounds.

By the early 1920's, wagon manufacturers had eliminated every size and style of vehicle not considered absolutely necessary. A two-horse wagon of a certain size skein, produced by the International Harvester Company, had once been available in 876 variations, each different in some manner. By 1922, the number of variations had been reduced to 16. With a standardized wagon, a farmer could build hay racks and other special bodies that necessity might demand, knowing that his equipment would fit any vehicle he might buy in the future.

Wagon makers exercised great care in the construction of a good farm wagon, particularly in the selection and handling of the wood stock. Lumber was air-seasoned from two to five years and, as a further guarantee against shrinkage, was stored in a dry room before being used. Gears were made exclusively of oak and hickory. Because of strength, oak was used for hubs, hounds, reach, bolsters and spokes. Because of resiliency, hickory was used for axles, singletrees, whiffletrees and neckyokes. Axles were hand-split, rather than sawed, to preserve the natural grain.

The generally poor road conditions of the early 1900's made the cost of transporting farm commodities exceedingly high. One estimate placed the over-all national cost at $450,000,000 a year. It was commonly believed that a system of improved roads throughout the nation could reduce this cost by about 60 percent. In 1915 the USDA estimated farmers were losing $250,000,000 annually, because of problems in getting products to market. Roads always seemed impassable when the grain market was at its peak. The protest that arose from these "mud-bound" conditions placed the farmer in full support of the "good roads movement" that was starting to sweep the country.

The farm wagon itself often was directly responsible for the breakdown of road surfaces. Tires 1½ to 2½ inches wide, standard on most earlier vehicles, had a tendency to cut deeply into soft earthen roadbeds. They also damaged macadam surfaces at certain times of the year. Several states adopted laws regulating the width of wagon tires while under heavy loads. In Ohio, the law prohibited the use of tires less than four inches wide on improved roads when loads of more than a ton were hauled, unless the surface was sufficiently dry or frozen. However, these laws were seldom enforced.

Not all wagons had high wheels. For those who preferred it, Studebaker manufactured this Low Wheel farm wagon. The hind wheels were ten inches lower than those on the regular style vehicle.

"Old Hickory" farm wagons were a product of the Kentucky Wagon Manufacturing Company. This Medium North wagon was regularly equipped with a boot end box, which was a handy attachment when shoveling grain.

99

The flaring grain tank box was used in large numbers in the Dakotas, Kansas and other wheat-growing sections of the country. Capacity was usually 100 bushels. The example shown was manufactured by the Peoria Wagon Company.

In the closing days of World War I, a matter involving farm wagons and other horse-drawn vehicles was drawing national attention. For generations wagons had been manufactured in two optional track widths: a "narrow track" of 54 inches, and a "wide track" of 60 inches (measured from center to center of tires). Although the northern states always had been partial to the narrow width and the southern states vice versa, this was not necessarily the rule. A farmer generally purchased the width most commonly used in his area.

When a farmer purchased an automobile he never gave wheel tread a thought. There was only one track available — 56 inches. By 1919 there were nearly 5,000,000 automobiles in the United States. The average auto was driven about 3,000 miles each year, compared to less than 300 miles for the farm wagon. Thus, the track width on nearly all country roads was controlled by the automobile.

During spring thaws and periods of rainy weather, on thousands of miles of unimproved roads, the constant stream of autos formed two deep wheel ruts, 56 inches apart, and threw up ridges on either side. Frequently, automobiles were equipped with chains which made the ruts still deeper.

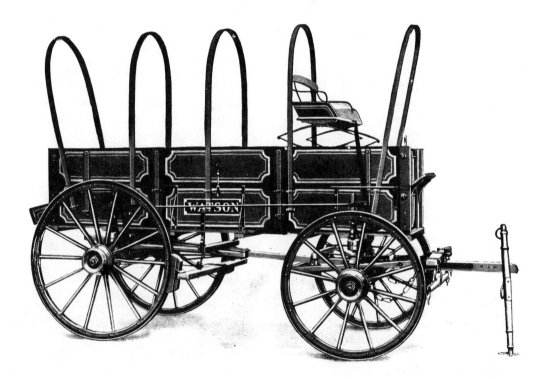

All of the wood in wagons manufactured for use in the Southwest was soaked in oil before painting as an aid in withstanding hot, dry climates. This Southwestern wagon was sold by Sears, Roebuck & Company. Top bows were furnished at extra charge.

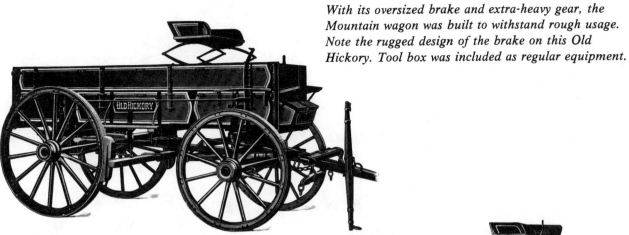

With its oversized brake and extra-heavy gear, the Mountain wagon was built to withstand rough usage. Note the rugged design of the brake on this Old Hickory. Tool box was included as regular equipment.

Another type of Mountain wagon, with a body style known as the "California rack bed," was popular in some sections of the West. This Studebaker had a box that measured 15½ feet long, with a capacity of 8,500 pounds.

Nearly every manufacturer of farm wagons produced a small, one-horse model. This Moline-Mandt had a capacity of up to 1,500 pounds.

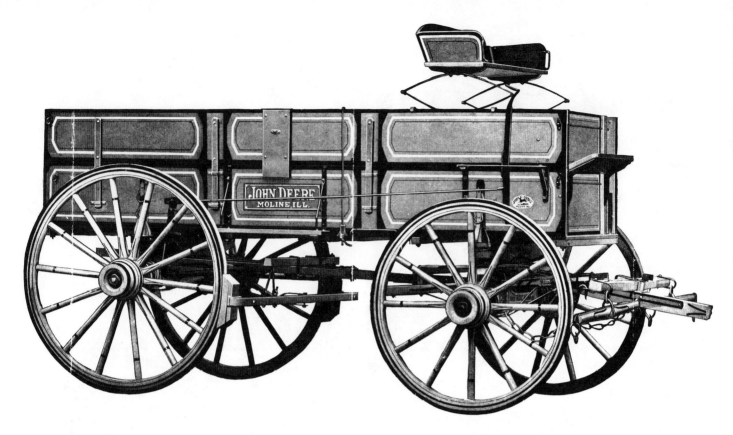

The 60-inch wide-track wagon overlapped the established road track by four inches. Consequently, the wheels on one side of the vehicle had to ride the ridge outside the rut. Often this ridge would break down, allowing a wheel to crowd into the rut, forcing the adjacent one out. If the wagon was heavily loaded, this additional strain on the axles and twist on the wheels occasionally broke one or the other.

The problem was serious and farmers demanded a solution. For the sake of efficiency and economy, the United States government recommended that the future track of farm wagons conform to the track of motor vehicles. A USDA report contended that a standardized track was certain to result in better road conditions, relief to teams, and lower wagon repair costs. Manufacturers soon were devoting considerable space in their advertising circulars to the merits of the new "auto track" wagon.

This John Deere of 1919 typifies the standardized farm wagon then appearing on the market. Although the demand for farm wagons gradually diminished in the years that followed, a few companies continued their manufacture through the 1930's.

For making hay or for hauling bundles, farm wagon gear such as the above was used. A rack, built according to the user's specifications, was attached to the gear.

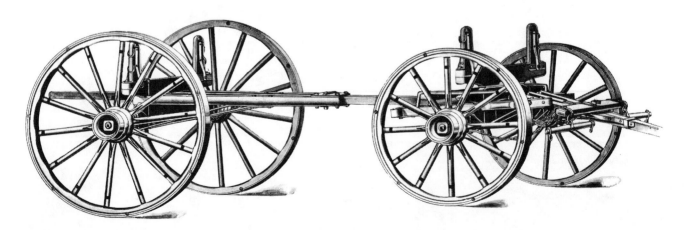

A wagon dealer in Texas noted in a letter to the **Tractor and Gas Engine Review**:

The automobiles and trucks are here to stay and the road and track is largely formed by them. Other wheeled vehicles must follow the track. It is no use to fight the inevitable. A fifty-six inch track beaten down and rolled out on a Texas road is and will be the track; you cannot change it, and if you stop and think about it awhile you don't want to. Is it not easier to follow the beaten track than to cut out a new one? If you are as far behind the times as I am and still drive a buggy, try a fifty-six inch one and see if you do not prefer it . . .

The International Harvester "Auto Wagon" was the first serious attempt by any manufacturer to produce a vehicle that could replace the horse for trips from farm to market. Pictured is one of the earliest IHC Auto Wagons, built about 1908. (International Harvester Co.)

The Avery Company in 1914 manufactured this "Country and Farm" truck, featuring special cast steel rim wheels which were designed for rural hauling. It had a carrying capacity of three tons.

As with the automobile, the motor truck found its first acceptance in urban areas. In the cities and larger towns the commercial truck and delivery wagon had appeared in sizable numbers by 1911, and it was considered only a matter of time until the gas-driven vehicle would supplant the horse. More than 10,000 trucks were manufactured that year, eclipsing in 12 months their entire production from 1903 to 1910.

An occasional model was designed for rural use, but several years would pass before the motor truck would be commonplace on the American farm. In addition to cost, which was considerable, the lack of good roads discouraged the extensive use of trucks beyond the city limits. But with highway improvements and the development of trucks and tires giving more reliable service under difficult hauling conditions, the number of these vehicles used in rural areas increased steadily.

Ransom E. Olds produced this Reo stake-body truck in 1912. It was rated at 12 horsepower and could haul a maximum load of 1,500 pounds. Price — $750.

104

In 1917, after two years of testing, the Ford Motor Company introduced the one-ton Model TT truck. Buyers could purchase the chassis and equip it with any type of body desired for their particular needs. The Model TT was especially popular with farmers.

Many of the earliest farm trucks served a double purpose. They could haul light loads of from one-half to three-quarters of a ton, and they could be converted into a vehicle for passenger service by adding an extra seat or two, thus taking the place of the much-used spring wagon.

A good example of this type of vehicle was the International Harvester "Auto Wagon," first manufactured in 1907. Powered by a two-cylinder, horizontal opposed engine, this air-cooled, chain-driven conveyance was designed to run at a maximum speed of 20 miles an hour. The ignition system featured a low-tension magneto, with a dry cell battery for starting. The truck's solid rubber tires, measuring up to 44 inches in diameter, provided the mobility required to travel the abominable roads of the period. Available in either 56-inch standard or 60-inch wide track, it offered a choice of track width, as did the earlier horse-drawn vehicles.

By 1915 an estimated 25,000 trucks were on American farms. The high-wheeled type vehicle had become outdated and was superseded by a truck with more conventional lines. There were those persons who formerly had envisioned a universal farm power machine that could plow the ground, plant the seed, harvest the crop, and be used as a touring car on Sundays to take the family to church. But the realization of this dream was becoming more faint, because the tendency was towards special machines adapted for special purposes.

The Samson truck was manufactured in Janesville, Wisconsin, by a division of General Motors. This 1919 Samson was priced at $995. "War tax" was extra.

FIRST – *The Telephone*
NEXT – *Rural Free Delivery*
NOW – **Muskegon Two-Ton Trucks**

Each in their turn have increased the efficiency of progressive farmers.

One farmer makes regular daily trips with his MUSKEGON. Not only to carry his own products, but his neighbors' as well. This express service has established a ready market, besides serving the farmers with a ready outlet for perishable products.

MUSKEGON TRUCKS more than pay their own way wherever used. Send for catalog. Don't put it off. Write for it today.

Address Dept. C.

Special proposition for dealers.

MUSKEGON ENGINE CO.
MUSKEGON, MICH.

This advertisement appeared in the April, 1919, Tractor and Gas Engine Review.

Production figures revealed that more than 125,000 new trucks rolled off assembly lines in 1917. In the city, trucks of various sizes were displacing horse-drawn vehicles for every type of hauling. The milk man, who followed his route in the early-morning hours, was about the only deliveryman still clinging to the horse, and even he was losing out. The smell of burning gasoline was becoming the order of the times.

The typical truck of 1917 had a four-cylinder engine. The cylinders were cast either in pairs or all in one block. It was water-cooled, using pump circulation through a cast frame radiator. Other features included electric starting and lighting, three-speed gear box, worm drive, dry-disk clutch, semi-elliptic springs, splash-system lubrication, and artillery-type wooden wheels.

The chassis was constructed with fewer refinements and with greater strength than found in automobiles. Pneumatic tires were used on light trucks where high speed was required, but tires of solid rubber were standard on larger vehicles.

Many farmers hesitated to buy a truck, even when they had considerable hauling to do. So a few enterprising men began establishing routes in different localities, operating one or more trucks between given points on a regular schedule. Between these points, they stopped to pick up whatever produce the farmers had to offer for transport to city markets. On the return trip, they delivered any supplies needed by the farmers.

Hostilities in Europe in 1914 created a critical shortage of horses. Soon practically every available horse on the continent was seized and placed in war service. A heavy toll of horseflesh resulted. It was estimated that the average animal survived only one week at the front. During the course of the conflict, various warring governments purchased thousands of horses and mules from the United States. The number totaled nearly 2,000,000 by Armistice Day. Both the scarcity and the price of these animals, which had been increasing for years, was certain to continue. Realizing this, many farmers now turned to trucks.

The truck also assumed a prominent role in the European war. Actually, the war acted as a sort of proving ground. When the big guns on both sides of the line were demolishing railroads and highways, the sturdy truck pulled its heavy load through the mud, climbed over obstructions, and carried supplies of munitions and food to waiting men. Trucks by the thousands, lining the roads like ants, hurried supplies to their destination. The picturesque army teamster was rapidly passing into history.

When the heavy demand for "Liberty Trucks" ceased in 1918, manufacturers began seeking markets that could absorb the large volume of vehicles they were geared to produce. The farm market, being virtually untapped, became the target of vigorous advertising. Although up-to-date hauling to most farmers meant a wagon with the new auto-track, many were trying the truck, and in most cases the reports were favorable.

Tabulations drawn during this transitional period revealed that hauling by horse and wagon cost 25 cents a ton for each mile traveled, while the expense of maintaining a horse was increasing at an annual rate of seven percent. The average farm haul was nine miles, requiring five hours for a two-horse team. The ever-increasing number of trucks sold during the 1920's reflected a growing public awareness of their advantages in economy and speed.

The Maxwell Motor Company, famous for its automobiles, for a time also manufactured a line of trucks. Shown is a circa 1920 Maxwell 1½-ton truck with stock-rack body.

The White Company of Cleveland, Ohio, built automobiles for several years but eventually switched to trucks. This farm marketing scene shows a two-ton White of about 1920.

Above: Graham Brothers trucks were manufactured in a variety of body styles and were equipped with engines supplied by the Dodge Brothers of automobile fame. This model was built about 1925.

Below: The Advance-Rumely Thresher Company produced this 1½- to 2-ton truck for several years in the 1920's. A price of $2,200 in 1925 did not include closed cab, pneumatic tires or electric starting and lighting.

108

FARM HAULING
by MOTOR TRUCK

FARMERS in every section are turning their hauling over to good motor trucks. There are plenty of practical reasons for the change.

The motor truck owner saves hundreds of hours per year and that means saving expensive labor for productive work at home. The automobile has shown him how to cut time and distance. Concrete and other hard roads are ready for his fast hauling. Distant markets are in easy reach. He is able to sell where prices favor him, and to buy likewise. In marketing his crops he cuts two days of old-style plodding down to a half day. Whatever he hauls—milk and cream, garden truck, live stock, fruit, grain and hay, sand, gravel, etc.— he does it *easier* and *cheaper* by motor truck.

In the full line of International Motor Trucks is a size and style to fit every farmer's needs. All sturdily built for hard, all-around farm use. Plenty of surplus power for heavy going.

The Speed Truck shown above carries a ton load easily and quickly. It is equipped with electric lights and starter, power tire pump, truck cord tires; special farm bodies available for your particular needs. The Speed Truck makes an excellent all-purpose truck for the average farmer. Where larger loads must be hauled, heavy-duty Internationals may be had for loads up to 10,000 pounds (max. cap.). Let us tell you how conveniently you can secure an International Motor Truck. A letter will bring complete information.

INTERNATIONAL HARVESTER COMPANY
OF AMERICA
(Incorporated)
606 So. Michigan Ave.   **Chicago, Ill.**

105 direct Company Branches serve International Motor Truck owners

INTERNATIONAL
TRUCKS

The transportation of any commodity is an important item in the cost of that product. Anything that could reduce the cost of moving farm produce would have a tendency to lower the cost of living. The truck was becoming an essential factor in providing this service. One man handling a steering wheel could do the work of three or four holding the reins.

Despite the truck's increasing popularity, horse-drawn vehicles were not yet obsolete. At corn-husking time, wagons were regarded as a necessity. For short-distance hauling, especially over bad roads, they held their own for several years. From the viewpoint of many farmers, the team and wagon were adapted to a larger number of uses under a greater variety of conditions. There also was "horse sense" on the part of the team — a desire and an ability to understand the wish of the driver, something not possible with any mechanically-driven vehicle.

The farmer who lived on an improved road usually favored the truck, particularly if he lived several miles from the nearest town. Until the advent of trucks, residing 10-15 miles from a good market was considered a serious disadvantage, unless a steam or electric railroad passed nearby. To cover these distances with a team and wagon required several hours, if not the entire day. With a truck, 20 miles was no greater than five miles had been with horse-drawn vehicles.

The advantages of motor-driven vehicles were most apparent on hot days. Many horses returned from market weakened, and sometimes were unable to make a trip the following day. The untiring truck was ready for duty the next morning no matter how late it got home.

Time was important in the business of dairying, because milk and cream, both perishable, could spoil during hot weather if not delivered to market promptly. Milk cans were often hauled by wagon to the railroad station where they stood unprotected from the sun until the train arrived. With a truck, the cans could be hauled directly to distributing centers in the city. Eggs, too, were delivered more frequently, minus the damage that occasionally resulted from a jolting wagon.

In addition to marketing light farm products, trucks transported grain and livestock to markets of the midwest. Hauling livestock by truck reduced shrinkage because it shortened the time the animals were in transit. By 1930, with about 800,000 trucks on American farms, cattlemen were shipping more than half of their produce in this manner.

The universal use of the motor truck in agriculture stimulated the growth of large urban areas while restricting development of smaller towns. It encouraged the growth of a few central market places and reduced others to positions of diminishing prominence. Thousands of towns, once the social and trade centers of their respective communities, became languid and faded into virtual obscurity.

AVERY COMPANY PEORIA, ILL., U.S.A.

Avery Motor Truck equipped with Grain Body and All-Weather Cab.

Grain Body—Inside dimensions 90 inches long, 50 inches wide by 26 inches high.

This six-cylinder Avery motor truck of the 1920's was equipped with a 50 by 90-inch grain body and an "all weather cab." It could haul a pay load of 1¼ tons capacity.

THE AMERICAN CHAMPION REVERSIBLE ROAD MACHINE.

By the 1890's, implements that were vastly more efficient than earlier hand methods were becoming available for building and maintaining roads. The American Road Machine Company claimed that its 1892 grader could do the work of 30 or 40 men.

6
Getting the Farmer Out of the Mud

The average country road of the 1890's had seen little improvement over the years. It often wandered randomly, with drainage taking its natural course, occasionally down the center of the lane. During late-winter thaws and extended spring rains, mud was axle deep. When dry weather prevailed, it choked in dust. Since maintenance funds were usually scarce, a country road was seldom graded more than once or twice a year.

Public roads were divided into two classes: those that were made of earth, known as unimproved; and those that were resurfaced with gravel or macadam, known as improved. The macadam road, which orginated in England in the early 19th century, consisted of crushed stone, from 1 to 2½ inches in diameter, spread in two courses on a solid and well-drained foundation. Each course was sprinkled with water, then compacted with a heavy roller. To fill the voids and firmly bind the material, a thin layer of stone "screenings" was added to the driving surface.

Only an occasional main artery in heavily-traveled sections was of the improved variety. Travel speed was still governed by the horse's gait, and the so-called permanent highway was not considered a necessity. The transportation system of the country was based on a complex system of railroad tracks, which totalled more than 200,000 miles in 1900.

The roller was deemed an indispensable implement in the work of road making. This was especially true in macadam construction, as it was essential that each course of stone received maximum compaction. The Galion steam road roller was one of the leaders in performing this service.

The Huber steam roller was introduced about 1905. Its return-flue boiler was similar to those used on the Huber line of steam traction engines. Note the scarifying attachment.

Another popular road roller was the Monarch. The model shown was manufactured in 10- and 12-ton sizes and featured a jacketed boiler, two speeds and a double-cylinder engine.

The new century heralded the arrival of the automobile and the beginning of the end for horse-dominated travel. With it, however, came numerous problems for the pioneer motorist. Unlike the buggy, the utility of the automobile depended almost entirely on the surface over which it was driven. Motorists unanimously condemned the highways of the period and suggested that a new concept of road building be adopted. But no person relied more on roads than did the American farmer, who also was beginning to petition for their improvement. A circular, issued at this time by the USDA, made the following observation:

> Life on a farm often becomes, as a result of "bottomless roads," isolated and barren of social enjoyments and pleasures, and country people in some communities suffer great disadvantages. Good roads, like good streets, make habitation among them more desirable; they economize time in the transportation of products and reduce wear and tear on horses, harness and vehicles. They raise the value of farm lands and tend to beautify the country through which they pass; they facilitate rural mail delivery and are a potent aid to education, religion and sociability . . .

In 1905 a special train toured 34 states on an educational crusade for the farmer — showing him how to organize his forces and secure effective road legislation. Conventions were held in many localities, and at each a local good roads association was formed. Every new member was asked to sign a petition to his state representative in the general assembly on behalf of better roads.

A new sentiment known as the "good roads movement" was beginning to sweep the country. By 1906, eight percent of the roads in the United States had received an application of some sort of hard-surfacing material. The cost of macadam roads generally ranged from $2,000 to $5,000 a mile, depending on the proximity of materials used. In addition, a few states were experimenting with brick and concrete construction.

Funds for maintaining highways in the early 1900's were often obtained by a "road tax" which was assessed farmers. In some localities, farmers were given the option of paying or "working out" these taxes. Those who chose the latter labored under the direction of an elected district supervisor. The inequality of service given to road work by both supervisor and taxpayer frequently condemned the system. If oat sowing or corn planting was in progress, the attitude was often one of "Well, we can work the roads when we can't plant." When field work was delayed because of wet weather, making help available for a day or two, the roads usually were too muddy to repair. A rural correspondent in the **Ashland Press** reported;

> There seems to be a disposition on the part of some people not to work their two days upon the public highway as the law demands, although they use the roads as much as others who work without complaint. We are glad to say that for this district we have but one shirker out of twenty-six men, and this one has neither horse nor buggy . . .

The spreader wagon was used universally in the construction of macadam roads. With it, material could be uniformly distributed over the surface of a roadbed to any desired depth. This Russell wagon had a capacity of 1½ yards.

Before it was rolled, macadam material was given an application of water with a sprinkling wagon. The average-size tank held about 600 gallons. Similar vehicles were used for distributing oil.

Grading work takes place on a road near Tuscola, Illinois, in 1913. Smaller graders used for heavy highway work could be operated with from four to six horses, while larger machines required an engine of at least 25 drawbar horsepower. The tractor used here is a 25-horse Minneapolis.

By 1909 the automobile had launched a world-wide crusade for dustless roads. While the annoyance of swirling dust had always been regarded as unavoidable, automobiles, often traveling at "breakneck" speeds, were now making action imperative. Conditions were intolerable for motorists, drivers of horse-drawn vehicles, and residents living along the routes. In a few instances, on roads where auto traffic was unusually heavy, the great clouds of dust became so dense that life was endangered. It was perilous for pedestrians to cross safely or for automobiles to pass each other.

State highway commissions in several parts of the country began experimental work with various types of petroleum products in an effort to find a solution. Treatment took different forms, including the use of crude oil and hot tar, which were distributed over the roadway with sprinkling wagons. These applications not only laid the dust, but also formed an impervious surface, binding together the material whether clay, sand, or gravel.

California, whose roads became particularly unbearable, pioneered an effort for their improvement. The need for settling dust had been recognized by state and local officials as early as 1898. If California had more than its share of dust, it fortunately produced the remedy in liberal quantities. Crude oil was available for as little as 60 cents a barrel, and in many localities there was no transportation charge.

By 1909 more than 1,000 miles of roads in the Golden State were oiled, and treated roads soon became more numerous than untreated ones. The rate of application ranged from 120 to 300 barrels a mile, applied to an 18-foot wide roadway.

What was termed as "road rescue" work was being carried on energetically in many of the eastern states. The macadam road, no matter how well constructed, was subject to grinding wear even under moderate usage. Few people realized that it lost about two inches of its surface each summer to dust that was created and carried off by the winds. With the use of oil or tar, the macadam road could be made to last almost indefinitely. Scarcely a state east of the Rocky Mountains had not at least some experimental roads treated with dust-laying material. The automobile was becoming an important factor in accelerating highway improvements.

By modern standards, the life expectancy of early pneumatic tires left much to be desired. Under favorable conditions they might yield 4,000 or 5,000 miles, but quite often they offered less. Although quality played no small part in their early demise, much of the blame could be placed on what was known as the "tracking habit."

From time immemorial, bad roads had been the rule. A buggy, when driven over these soft, earthen surfaces, invariably formed a well-rutted track. The drivers who followed habitually used the same track, with the narrow steel tires of each vehicle cutting it even deeper.

The arrival of the automobile did not change driving habits. Rubber tires and increased speeds removed all the loose dirt and pebbles from axle-deep ruts, leaving only the larger and often sharp-cornered rocks. Following in the same track seldom damaged steel tires, but the practice obviously caused many a delay for the early motorist. Some advocates of good roads recommended wider roadways with the thought that this would reduce the tracking habit. Others were not so sure.

For generations the farmer received his mail after much loss of time, travel over poor roads, exposure to storms, and other inconveniences that often made his lot seem a hard one. The trip into town by top buggy or spring wagon was made routinely two or three times each week. The farmer who procured his mail daily had to either farm by proxy or live in the shadow of the post office.

In 1896 Congress appropriated funds to provide mail delivery to rural areas. This service, Rural Free Delivery, began the following year on a trial basis, with 83 routes serving about 10,000 people. Its success was instantaneous. Appropriations for its operation increased from $40,000 in 1897 to nearly $13,000,000 in 1904.

Township officers in Worth County, Iowa, observe a road-building operation in 1915. The "workhorse" is a Hart-Parr "Oil King."

Rural routes were established in response to signed petitions. A majority of the farmers on a proposed route had to sign this written request, which was then forwarded to the Post Office Department. No route could be established on which fewer than 100 families lived or which was less than 20 miles long. With the endurance of the carrier's horse in mind, the average mail route for the country was held to about 25 miles.

This new service kindled far-reaching enthusiasm throughout rural communities. A farmer reported in the **Ashland Press** in 1901:

> Rural free mail delivery has begun and everybody on the route is on the mountain top of ecstasy. Farmers should have had this service long ago in consideraton that they have always been the standby of the government, though they are the last ones to get the comforts and conveniences due them.

Rural mail delivery, however, brought an air of apprehension to merchants of small villages who depended upon rural trade. With mail delivered to their doorsteps, farmers would have less occasion to visit these villages. Most storekeepers associated this with a decline in business. The fourth-class country post office — a frequent adjunct of the general store — was being closed down in many localities.

In addition to putting farmers in more intimate touch with mundane affairs, Rural Free Delivery was an important factor in securing better roads. To be accepted for rural mail service, a proposed route had to meet government standards. An inspection tour was made to deter-

mine if its condition would allow daily passage of the carrier's wagon or automobile. Once instituted, a route had to be properly maintained or the benefits of rural delivery could be withdrawn.

Each passing year saw the establishment of new routes and the appointment of new carriers. By 1912 Rural Free Delivery had 42,000 routes which served more than 20,000,000 people. The service had been extended to practically every farming community in the country.

By 1913 many states were introducing legislation providing for a state highway commission and some form of state aid. In Wisconsin, the state assumed one-third the costs of new highways and 20 percent of the costs of bridges, granting that state funds were sufficient. Counties and municipalities shared the remaining amount equally. This legislation was directed toward the construction of good highways connecting all leading trade centers and the eventual linking of adequate laterals to every rural community.

"Good roads days," declared by some governors, were becoming the vogue. In such states the entire population was asked to contribute its time to the betterment of local roads. Such a day, observed in Illinois, suggested the following: "haul stone, haul sand, fill low places, cut down humps, clean ditches, grade the roads, drag the roads, and cut the brush."

Although good roads days focused public attention on a universal problem, they often failed to provide any real improvement. In many instances this was attributed to a lack of skilled supervision.

A Case 30-60 tractor does double duty in 1915. The graders also are of Case manufacture.

By 1915 the trend towards light-weight general-purpose tractors for farm use was reducing sales of larger models. A new branch of service was opening up, however, for which these lumbering giants were well adapted — building and maintaining roads. The National Highway plan was to improve about 50,000 miles of trunk roads permanently and maintain the feeders in condition to allow year-round traffic. The gas tractor was predominantly "the thing" for performing this kind of service. Many tractors were purchased solely for road work, not only by counties and townships but also by private individuals. A few farmers leased their tractors for work on roads and found this practice quite profitable.

Properly maintaining an earthen road was not always an easy matter; at certain times of year it bordered on the impossible. Under favorable circumstances, however, the dirt road was considered a better surface for horse traffic and was not always objectionable from the standpoint of motor traffic. The key to a dirt road's maintenance was constant attention and the judicious use of two implements — the grader and the drag.

Water was the worst enemy of the earth road, and a well-crowned surface was essential for good drainage. A "patrol" grader provided this service and also kept the ditches open. Persistent use of a drag, particularly after light rainfalls, was necessary to keep the surface free from holes and ruts.

The Adams "Road Patrol" was a popular one-man, two-horse road maintenance machine.

*The Linn track-laying type tractor was an unusual-
looking machine. Here it hauls a heavy road grader
near Morris, New York.*

In 1911 President Hooper of the American Automobile
Association predicted that it would be only a matter of
time until the government would cooperate actively with
the states in building a nationwide system of public
highways. In July, 1916, President Woodrow Wilson
signed the Federal Aid Road Act, which provided this
assistance with matching funds of up to 50 percent. This
reflected the federal government's interest in adequate
roads for mail delivery, national defense and interstate
commerce.

The one-room district school was an institution
throughout rural America for many decades. Its origin
came at a time when families were large, roads were new,
and education, beyond the simplest rudiments, more of a
luxury than a necessity. The resourceful teacher had to
divide her attention among eight or more grades and provide instruction in everything from the alphabet to the
most advanced part of the curriculum.

By World War I, the familiar tolling of the district
school bell was no longer heard in some communities.
Scattered over the country were "little red schoolhouses"
with weeds growing in the dooryards and "closed" signs
nailed on their doors.

After years of argument and counter-argument the
centralized system was beginning to be adopted, principally because it afforded better instruction in individual subjects. At township debates farmers had often
opposed the issue, contending that "the old system is
good enough." Proponents quickly countered with the
query of why farmers did not then harvest with the cradle
and thresh with the flail as had their grandfathers.

When a township voted to abolish its one-room style of
education it almost always brought a clamor for better
roads, as children could no longer wend their way to
school on foot but had to be transported by wagon or
bus.

In 1912 a man from Indiana, Carl Fisher, had a unique idea for advancing the country's highway system.
There were about 2,000,000 miles of unimproved roadways in the United States. Every year a certain number of
miles were improved, a few miles here, a few there —
each section beginning nowhere and ending nowhere.
Political differences and local self-interest had held back
any real national progress. A patriotic impulse was needed.

Fisher's inspiration was to furnish that impulse and
provide a shining example of what could be achieved by
united effort — a highway thousands of miles long,
reaching from the Atlantic to the Pacific, with every mile
graded and hard surfaced. It would serve as a model for
additional arteries certain to be built in years to come.

Fisher reasoned that if only a portion of the country's
total road funds were centered on one highway, and if, in
addition, a public-spirited citizenry would actively support it, the project could be completed in a surprisingly
short time.

Working with a group of supporters, Fisher helped
devise a plan to finance the project. An association was
formed and incorporated, engineers and local
automobile clubs consulted, specifications and route
chosen. The aim of the association was:

The scarifier often was used for reshaping macadam and earth roadways. Its huge teeth served to loosen the road's surface, which was then regraded and rolled. The machine pictured was manufactured by the Russell Grader Company.

. . . to immediately promote and procure the establishment of a continuous improved highway from the Atlantic to the Pacific, open to lawful traffic of all descriptions, without toll charges; and to be of concrete wherever practicable. The highway is to be known as "The Lincoln Highway" in memory of Abraham Lincoln.

Ten million dollars was to be raised by popular subscription. The remainder needed, or about 60 percent of the total, was to come in the form of aid from states, counties, and municipalities along the route. The small expenses of the association were borne in part by the group of original founders and by the sale of maps, pennants, and other material.

Individuals from across the country pledged their support and contributed money. Prominent manufacturers were generous, many donating from $50,000 to $300,000. More than a hundred business concerns between New York and San Francisco pledged in writing to give an amount equal to one-third of one percent of their gross sales for three years. A man in Los Angeles donated one drinking fountain for each 10 miles of the Lincoln Highway in Illinois, as a memorial to his mother.

Construction on this monumental undertaking began in the spring of 1914. In a few weeks, the entire route from coast to coast was clearly designated with the red, white, and blue markers of the association. The labor, as well as the material for this preliminary work, was contributed by each locality.

Although work on the Lincoln Highway did not progress as fast as was originally planned, it was nearly comleted by 1919. Cross-country runs by auto and truck were beginning to be made at this time. Carl Fisher's dream was finally being realized.

In 1919 the United States embarked upon the most remarkable plan of highway construction in the history of the world. Vast expenditures were expected to result in the creation of a network of first-class roads connecting all communities and open up new avenues of trade for farmer and city merchant alike.

For the first six months of 1919 the Federal Aid Bureau, in charge of the disbursement of federal road funds, approved a total of 1,319 projects, involving the construction of 12,790 miles of public highways at a total cost of $133,833,300. Of this amount, $54,763,957 was federal aid. Additional projects were added at the rate of $25,000,000 each month.

This road work was carried out under the direct control and supervision of various state highway commissions. The federal agents merely coordinated the projects and made certain government funds were applied toward highways that had value as post roads for mail delivery and military purposes. Before federal money was available in a particular district, the state or county had to appropriate at least an equal amount.

California, which already had a commendable system of main highways, voted for the issuance of $40,000,000 in bonds to further extend its laterals. When completed, every town of consequence would be linked with the main trunk highways that crossed the state. In addition, a million dollar compaign was launched to secure funds for a highway into Yosemite Valley. Every motorist was asked to purchase a $5 coupon, with the proceeds to be used for the Yosemite Highway. These coupons soon became as conspicuous on the automobiles of that state as were the license plates.

What was happening in California was taking place nearly everywhere. In fact, with so much money becoming available in so many places, it was almost impossible to secure personnel with the necessary experience and technical skill to properly execute the work. Federal officers weighed the possibility of holding up a portion of government funds unless a workable solution could be found.

120

Road building goes on in the early 1920's near Cleveland, Ohio. The truck is a White.

Opposite above: A well-maintained country road about 1918. Note the horse and carriage in background.

Opposite Below: Sometimes local conditions made road excavation with steamshovels desirable. This steam-shovel is a Marion-Osgood. The location is near Columbus, Ohio. (Ohio Historical Society)

During the 1920's the good roads movement, with its adopted slogan "Get the farmers out of the mud," was in full swing. Abundant road funds and advanced construction technology yielded tangible benefits to every rural community. Macadamized roadways became common almost everywhere. More than 31,000 miles of concrete roads were in use by 1924. Motoring to town over "the pike" was becoming a new and exhilarating experience for many farm families.

The preparation of a roadway for concrete and other improved surfaces often entailed excavation and embankment to modify steep grades and elevate hollows. Horse-drawn, hand-operated drag scrapers and their larger cousins, wheel scrapers, were used on small jobs and on short hauls. These diminutive earth movers usually required the assistance of a road plow, working in advance to loosen the soil.

Larger jobs required the use of an elevating grader, working in conjunction with several horse-drawn dump wagons. This machine, a combination of plow and elevator, turned a furrow directly onto an endless belt-conveyor, which loaded the wagons as they were driven alongside.

In practical use, it was not uncommon to load more than 1,000 cubic yards of material in a 10-hour day. Although generally drawn by a large tractor, teams of 16 horses (or mules) occasionally supplied the power. Dump wagons, usually of 1½-yard capacity, were used in gangs. The number of wagons depended on the length of the haul. Five wagons were recommended for a haul of 300 feet, plus an additional one for each 100 feet thereafter. Thus for a haul of 1,000 feet, 12 wagons were needed.

In the years following World War I, the track-type tractor emerged on the construction scene to challenge the conventional wheel tractor. The track-type tractor provided greater surface contact. Its treads exerted less pressure per square inch, reducing ground displacement and slippage. In addition to improved mobility, the machine featured individual track clutches which reduced the radius required for turning. When one clutch was released, the power applied through the other clutch caused the tractor to pivot in a circle. This new trend in tractive power became firmly established after the formation of the Caterpillar Tractor Company in 1925.

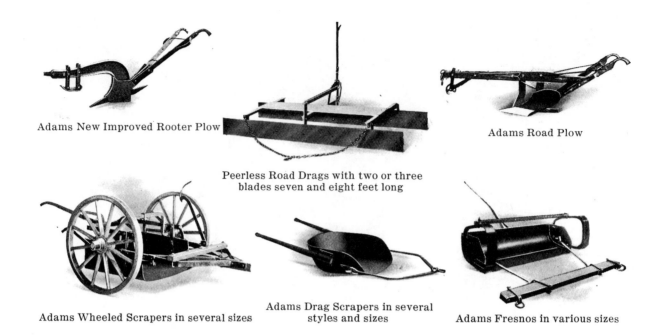

Adams New Improved Rooter Plow

Peerless Road Drags with two or three
blades seven and eight feet long

Adams Road Plow

Adams Wheeled Scrapers in several sizes

Adams Drag Scrapers in several
styles and sizes

Adams Fresnos in various sizes

Above: Light horse-drawn implements were popular on small road jobs. These were manufactured by the J. D. Adams Company.

Below: Dump wagons had bottom doors which were tripped by a hand or foot lever. The load was then discharged by gravity. This 1½-yard "Light Township" dump wagon was built by the Acme Road Machinery- Company.

The Austin "New Era" elevating grader was designed especially for high-powered crawler-type tractors. The use of a power take-off shaft greatly increased the capacity of the machine.

Introduction of the motor grader in the mid 1920's transformed road maintenance and subgrading into a one-man operation. The efficiency of this machine was enhanced by leaning front wheels, which reduced side-draft. The earliest graders were equipped with solid rubber tires, and the same engines used in medium-sized farm tractors. Introduction of the motor grader and the track-type tractor hastened a new era in highway construction.

Many benefits followed in the wake of better roads. As distance barriers vanished, the subtle temptation to forsake the old farm home left the minds of many young folks. When asked how far he lived from town, the farmer no longer grumbled "Four miles when the roads are dry, and 20 when it rains!" Prosperity was a particular "old duffer" who never relished being stuck in the mud.

An elevating grader and dump wagons in operation about 1920. The tractor is a Twin City 40-65.

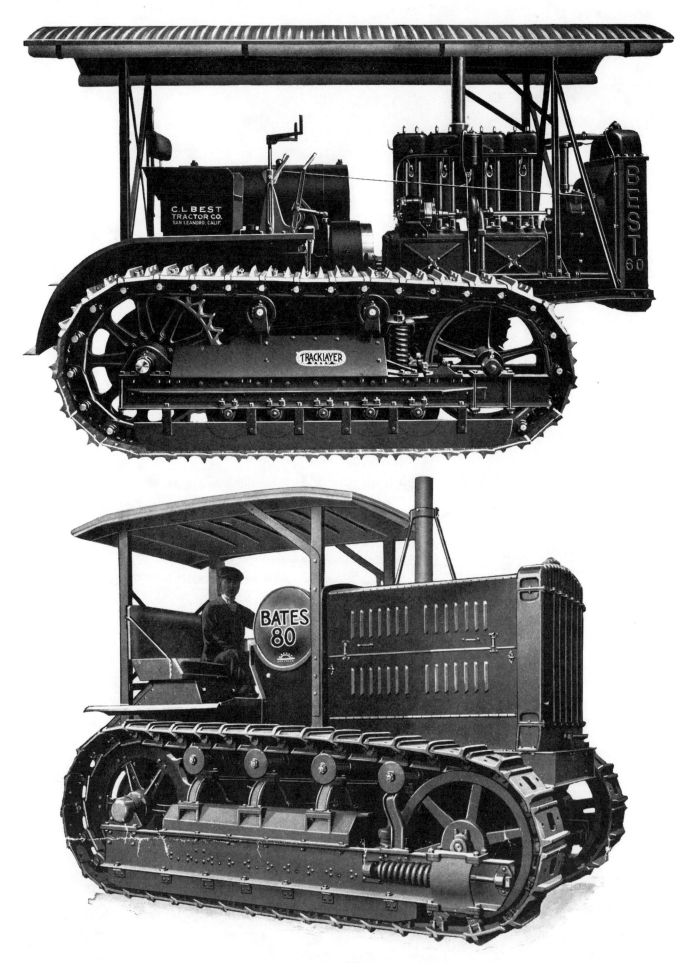

"Down in Georgia"

10-Ton hauling from steam shovel

Building good roads in the Texas Panhandle

These scenes are from a page of the 1922 Holt tractor catalog. The Five-Ton model is shown above. Hard at work in the center and bottom scenes is the heavy-duty Ten-Ton model. In 1925 the Best and Holt companies merged and formed the Caterpillar Tractor Company.

Opposite Above: The work of building roads required plenty of mobile power. Whatever the job, this Best "Sixty" crawler-type tractor of 1923 was equal to the task.

Opposite Below: The Bates "80" was the big brute of the "Steele Mule" line of tractors in the 1920's. It could handle a 14-foot road grader or the largest elevating grader with power to spare. It weighed 22,500 pounds.

Introduced in 1925, the Austin-Western was the first really successful motor grader on the market. The machine derived its power from a Fordson tractor engine. Note the leaning front wheels.

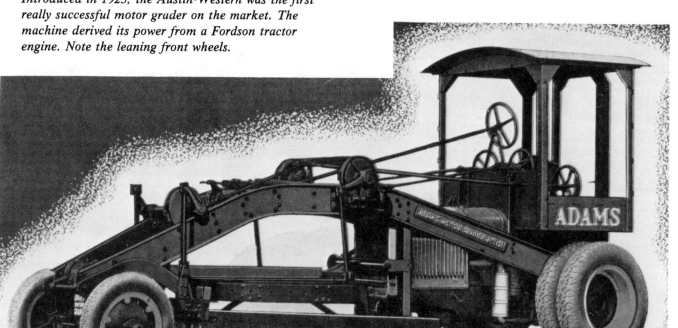

In 1931, Adams motor graders were equipped with either McCormick-Deering 10-20, Allis-Chalmers Model "U" or Case Model "L1" engines. Both solid rubber and pneumatic tires were available.

126

7

The "Devil Wagon" Invades the Farm

This typical piano-box buggy was popular at the turn of the century. Nearly every farmer owned one of these vehicles for visiting neighbors and making routine trips to town and post office. Color of body was usually black, with gear in Brewster green.

In the closing years of the 19th century the horse was King of the Road — as he had been for countless generations. As a drawer of the carriage, well groomed, well harnessed, and the pride of a considerate owner, his position in the American way of life was well established. Cherished for his companionship and his devotion to man's needs, he was considered an inseparable member of the household he served.

Anyone who didn't enjoy a ride behind a fine horse lacked something in his make-up. When a young man "went a courting," he first washed the best buggy, oiled the best harness, and groomed the best mare. Then he took his best girl for a ride to demonstrate what fast driving was really like. Most likely he passed everything on the road and kicked up enough dust to hide the landscape.

To supply the transportation needs of the country, hundreds of manufacturers were turning out a million horse-drawn carriages annually. With 60 factories, nearly 10,000 employees, and a total weekly production of 2,500 vehicles, the Queen City of Cincinnati was the center of the carriage industry.

Although one-seated carriages were manufactured in a number of fashions, ranging from the inexpensive runabout to the exquisite phaeton, the unadorned "top buggy" was most frequently seen in the farmer's carriage shed. Two distinctive body styles were available: the corning and the piano box. Because it was lower and easier of access, the corning was preferred by many businessmen, physicians, and drummers. But the piano box body was regarded with peculiar favor and was the predominate choice of those living in rural areas.

The corning-style buggy
was easy to get in and out of. Being
low in front and high in back, it
was a compromise between the
piano-box buggy and the phaeton.
Note the elliptic springs and "Sarven
patent" iron hubs, standard on the
running gear of most carriages.

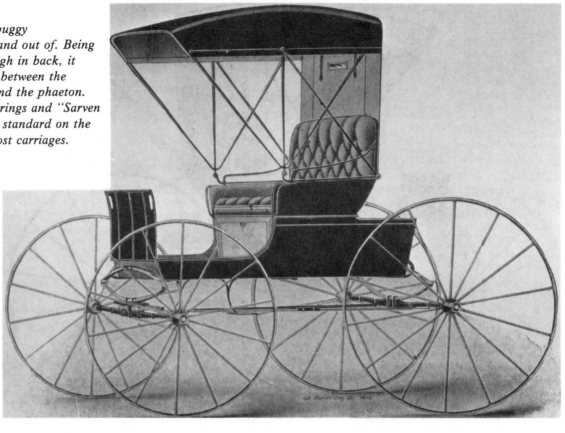

About 1908 the auto-seat
buggy was introduced. The seat
style of this vehicle was similar
to that used on early automobiles. An auto-seat buggy,
equipped with solid-rubber tires
(introduced in 1894), was considered a really
up-to-date "turnout."

The storm buggy made its appearance in the early 1900's. Compared with the common top buggy of the period, it afforded excellent protection when traveling in cold or stormy weather.

The extension-top surrey was a popular family carriage. Curtains protected the occupants in cold weather, and in summer the top could be folded back. Surries and buggies were manufactured with either rubber tops, full leather tops, or a combination of the two, known as leather quarter tops.

The canopy-top surrey was a handsome vehicle in which families took great pride in riding. It was just the thing to show off a good spirited team of horses. The top could be removed easily when an open carriage was desired.

The surrey was a capacious, two-seated carriage built to accommodate four or more persons. It was commonly used to take the farmer and his family to church on Sundays. Unlike the buggy and most other passenger vehicles, its top was available in two different styles. The extension top was characteristic of those used on early automobiles and could be folded down when the weather was favorable. The canopy-top vehicle, celebrated for its stylishness, was sometimes known as "the surrey with the fringe on top."

The runabout was essentially a piano-box buggy without a top. It was designed to meet the requirements of those desiring a serviceable vehicle at an inexpensive price.

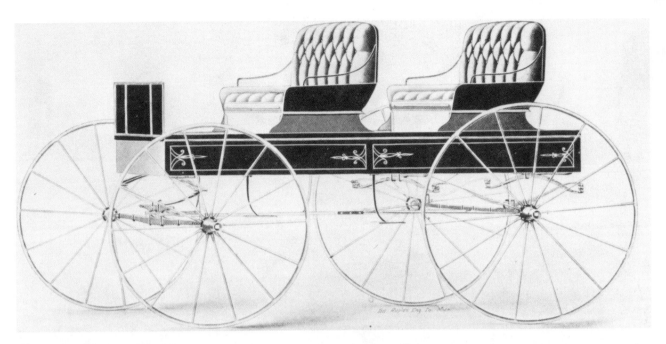

The spring wagon was a general utility vehicle used for hauling all types of loads and on all kinds of roads. Like farm wagons, spring wagons and carriages were manufactured with a choice of track widths.

The spring wagon, occasionally referred to as "the poor man's surrey," was a light and durable, two-seated vehicle that was also quite popular among farmers. Although essentially a family conveyance, its seats could be removed for hauling light loads. Carrying capacities ranged from 750 to 1,500 pounds. When fitted with a canopy top, it made a nice "outing" wagon.

Sleighs often were used for traveling in winter, whenever the roads were snow-covered. "Family sleighs," with a removable rear seat, were favored by many farmers, because they were suitable for either pleasure driving or light hauling. Heavy hauling jobs required the services of two horses and a bobsled. Any wagon could be converted into a bobsled by removing the box and attaching it to a set of "bobs."

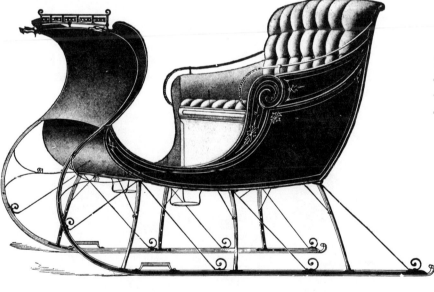

The Portland Cutter was unrivaled as a vehicle for winter sleighing. Buyers could choose from many elegant styles.

The bobsled was manufactured in several styles for teaming and farm work. The example pictured had a carrying capacity of 6,000 pounds.

Rambler
Automobiles

Model Forty-one
Specifications

BODY—Detachable tonneau, capacity five or two passengers.

WHEEL BASE—106 inches.

FRAME—Pressed steel throughout.

WHEELS—Wood, artillery type, 34 inches in diameter.

TIRES—Goodrich Quick Detachable, 4 inches, options G. & J. or Hartford-Dunlop.

MOTOR—Double opposed, entirely enclosed with transmission gear, making unit power plant.

POWER—22 horse power.

GASOLINE CAPACITY—13 gallons.

COOLING—Water, thermo-syphon system.

LUBRICATION—Mechanical force feed oiler.

IGNITION—Jump spark.

BATTERY—6-volt, 60-ampere storage.

TRANSMISSION—Planetary, two speeds forward and reverse.

CLUTCH—Multiple disc.

FINAL DRIVE—Single chain.

EQUIPMENT—Two gas head lights with large separate generator, oil side and tail lights, exhaust alarm, pump, tools and tire kit.

TOP—$75.

PRICE AS ABOVE—$1350. With torpedo deck in place of tonneau, otherwise as above, $1250; torpedo deck as an extra to touring car, $25.

The Rambler was becoming popular with farmers in 1909.

132

The International Harvester "Auto Buggy" was one of the more popular high-wheelers of its day. According to the IHC catalog, this vehicle could "travel long distances without delays for resting and feeding, without liability of a runaway or excessive urging to keep it moving." As shown here, the Auto Buggy frequently displaced the surrey as a family vehicle. (International Harvester Co.)

The dawn of the 20th century saw a totally new concept in transportation make its debut on the American scene. It introduced a craze that would soon affect much of the populace and eventually antiquate the time-honored horse and buggy. But in the beginning, the horseless carriage did not excite universal admiration.

For several years the farmer assailed these vehicles, and those who operated them, with withering denunciation. He often vented his displeasure by referring to them as "devil wagons," "infernal machines," or "toys for the rich." Sometimes he carried a shotgun in the back of his buggy or spring wagon, ready to administer justice to this newly-made "curse of the highway."

The farmer's animosity toward the automobile was not entirely unwarranted. Horses, unaccustomed to this new machine, were alarmed by its disquieting noise and alien appearance. Even the least excitable animals often became unstrung when an auto, "going like sixty," passed with a rattle and snort. Acting as if they owned the road, a few audacious motorists resorted to hornblowing and swearing when overtaking buggies and other slow-moving vehicles. This often tended to make horses unmanageable. But the act that most exceeded the bounds of propriety was opening a muffler cutout while under the very nose of a horse. Such an act invariably precipitated a break for the ditch. The motorist who lacked the good sense to recognize the rights of the horseman created a prejudice that was not soon forgotten.

Motorists were exhorted to be courteous to drivers of horses on public highways. When a farmer "turned out" in an accommodating manner upon meeting an automobile or gave part of the road to allow one to pass from the rear, the polite motorist thanked him for his graciousness. An amiable disposition was not without its rewards, as one pioneer autoist of 1908 related in the **Gas Review**:

I have gone over the same course taken by a friend of mine whom a farmer threatened to shoot if he ever came in that vicinity again, and by practicing common courtesy have had the farmers throw open their barns for me to store my machine and make necessary repairs; they even assisted in the work and were glad to help.

In the earliest stages of the industry, nearly everyone believed the automobile was a luxury which would appeal chiefly, if not exclusively, to the wealthier class for pleasure purposes. It also was assumed that few persons could ever learn to operate one. Any farmer so flagrantly bold as to consider owning or driving such a contraption was regarded as a shiftless old sport, sure to ruin himself and his family.

In the days prior to about 1906, when the populace was divided into two factions on the automobile question — those few city dwellers who used and advocated them and all the rural residents who cursed them — a few brave spirits dared to predict in the course of time the final "passing of the horse" in the big cities. But none

was so rash as to say that before the horse became obsolete in the cities he would give way to the automobile on
the farm; or that thousands of farmers would themselves
own these vehicles and jeer their neighbors whose skittish
two-year-olds pranced during a chance meeting on the
highway.

By 1909 an incredulous change in sentiment was taking place. As measured by the passage of time it had
winged its way almost overnight, but as measured by attitude it seemed to belong to another generation. The
automobile had reached a stage of development where it
was no longer considered a device to satisfy the man with
sporting tendencies, but rather was a necessity of life. Its
increasing reliability was attracting the attention of those
who did not expect to lay out hard-earned money unless
full value in service and satisfaction was to be obtained.

Perhaps no person was as practical as the American
farmer, who was beginning to realize just how valuable
the automobile could be. After all, if it was beneficial in
the cities where distances were relatively short, its
usefulness should be multiplied in the country where
homes were widely separated and remote from churches,
stores, and other commercial establishments.

The automobile as a means of saving time was beginning to have wide appeal. There were certain periods
when farm work had to be completed without delay or
the results of a season's labor could be partly or entirely
lost. During the harvest season, the adage that "time is
money" was literally true. A breakdown of machinery
that resulted in a trip to town represented a serious loss,
unless the trip could be made quickly. This is where the
auto could fully demonstrate its worth.

*From 1905 to 1910, Sears, Roebuck & Company
manufactured motor buggies. This 1910 Model J, complete with fenders, top, side curtains, and "a cozy
storm front for use in rainy or cold weather," sold for
$410.*

There also were other demands for quick trips between
town and farm. The automobile, sometimes called the
"annihilator of distance," was a boon to humanity in
that it enabled the physician to reach his rural patients
faster in times of accident or sudden illness.

Hundreds of farmers replaced their buggies with new
automobiles, and thousands debated the question of
whether to purchase. Manufacturers, aware of the op-
portunity, began building vehicles especially adapted to
agricultural needs. These "farmer's cars" were constructed, for the most part, with a detachable tonneau
and top. Produce and supplies could be hauled during
the week, and on Sundays the top could be put on for
pleasure driving.

Thomas B. Jeffery of Kenosha, Wisconsin, was one of
the first to anticipate this new market. In 1908 he began
advertising a two-cylinder "Rambler" that could be converted into a general utility machine in just a few
minutes. At the pleasure of the operator, the vehicle
could be used as either a five-passenger touring car, a
two-passenger roadster, or, by omission of both tonneau
or torpedo deck, a platform for carrying merchandise.
With crates of eggs or milk cans on the rear deck, the
operator could make a flying trip to market or creamery.
Or his load might consist of a calf for the butcher or
harness to be repaired. The horses could remain in the
field mowing or raking.

The high-wheeled motor buggy, sometimes called the "gas wagon", attracted not a little attention and favorable comment. Costing about half that of its more fashionable prototype, it placed the advantages of gasoline within the reach of those of moderate means. The motor was carried above the axle on wheels ranging from 36 to 50 inches in diameter, giving the greatest possible clearance over stones, stumps and ruts.

Striking out at more of an angle, these high wheels effectively reduced vibrations from unevenness in the roadbed. Wedge-shaped, hard rubber tires were standard equipment. Usually they outlasted the ordinary inflated tire by 100 percent. The transmission had a peculiar advantage in that it would slip on hard pulls just before the strain reached the breaking point. As these vehicles were operated with only one lever, they were easy to control. And because there were no foot levers, they were nice in cold weather when a lap robe was needed.

The buggy-type automobile was seldom seen in large cities, but in rural areas where it was often almost impossible to travel the roads, it was considered by many to be indispensable. Nowhere was this more evident than in the West, where there were hundreds of miles of roads

Another high-wheeled vehicle of the period was the "Buggycar," manufactured in Cincinnati, Ohio. Its air-cooled, 2-cylinder engine developed 18 horsepower. Price was $585.

over which the conventional auto could pass only under the most favorable conditions. Many streams had no bridges and had to be forded. In truth, the abominable highways of the West necessitated an automobile built to fit the conditions or its use abandoned altogether. The motor buggy was used successfully in scores of places that would make most motorists turn white — if not turn back.

One disadvantage of the motor buggy was that it provided less speed, usually not more than 15 miles an hour for the comfort of its occupants. It was not as luxurious as the touring car and the owner was not as distinguished as the man who rode the Pullman. But for the man who wanted to save money for the "rainy day," it supplanted the horse and carriage nicely.

Caron

The Ford "Model T" was introduced in the autumn of 1908, and during the first year 10,607 units were assembled. This was by far the largest number of cars ever turned out in a 12-month period by any manufacturer. Henry Ford saw clearly that the greatest possible success in the automobile field would be achieved by the manufacturer who offered an inexpensive vehicle — one that every man who could afford a horse and buggy could buy. Ford cars were built economically with no extra trimmings, no frills or fixings. They were plain, unpretentious vehicles designed for the workaday world. The "Tin Lizzie" was highly instrumental in overcoming the prejudices of the American farmer against "newfangled machines."

The automobile was beginning to alleviate the humdrum of rural life. So common was the auto becoming that the agricultural college curriculum in several states included a short course in its use on the farm. Vehicles were used in demonstrations, and classes were instructed in their operation and maintenance. Three years earlier the automobile had been virtually unknown in rural districts, and farmers still chafed when clattering machines from the cities frightened their horses, but now the farmer was himself joining the speed-mad class.

The Auburn, first manufactured in 1902, was highly respected within the automobile industry. This 1909 model featured a 2-cylinder, horizontally-opposed engine that developed 24 horsepower. Price, without top, was $1,250.

Seeing city men whizzing by their places at 30 miles per hour, many farmers often had imagined that the automobile was good for nothing but insane speed. One fellow appeared at a Kansas City dealership and applied for a car that would run slowly. "My family will not stand for riding fast!" he explained. He bought one after a demonstration and a short time later boasted that he had driven the 10 miles from his home to the nearest town in just 20 minutes.

In Kansas, a robber had been frequenting the farmsteads of a rural community for a fortnight. One morning an early-rising farmer saw the thief leaving his stable just before dawn. He immediately telephoned his neighbors and a pursuit by automobiles was organized. In the meantime the robber got an hour's start, but was caught six miles away sitting atop a water tank while waiting on a train. Farmers who once abhorred speed were now finding it an asset.

In its first full year of production, 1910, the Hudson became a highly esteemed automobile. The stylish touring car of that year was priced at $1,150.

Of the several air-cooled automobiles once built, the Franklin was most successful. According to its manufacturer, the use of air cooling eliminated 177 water-cooling parts. This 1909 Type G touring car had a price tag of $1,850. Top was extra.

The Self-Starting Winton Six-Teen-Six

The Winton was perhaps the most successful of the more than 80 automobiles once manufactured in Cleveland, Ohio. The group seated in this 1910 Winton "Six-Teen-Six" apparently were about to embark on an outing.

What about the horse — good Old Dobbin? Well, who supposed that his day was not already long enough and filled with sufficient tasks? Nothing was more wearisome on a draft animal than to be driven over a hard road after a tiring day in the field. The automobile not only reduced the number of horses required on the farm, but it also provided relief for those engaged in field work during the rush season. The farmer could eat a leisurely dinner and then, with merely the turn of a crank, roll into town in 15 minutes or a half hour instead of spending three or four times that long jogging monotonously in a buggy. One farmer said he motored to town, a distance of four miles, and conducted his business at noon while his horses were resting in their stalls. The trip was relaxing and the time consumed so brief that he was able to send the team to the field at the usual hour to resume the day's work.

The horse was a noble animal and a good servant, but he also had his limitations and his faults. The faults were occasionally those of temperament, sparked by a glimmer of intelligence which ranged from a keen sense of humor to downright cussedness.

This fact could be appreciated by anyone who had ever tried to catch a horse in a pasture — one that would stand with a twinkle in his eye until a hand was about to be laid upon him, then whisk away a few rods and wait expectantly to repeat the joke. All of the idle talk about the horse being man's best friend didn't make much of a hit after 14 times around the pasture with a lump of sugar in one hand and a halter in the other.

When a farmer purchased his first automobile, it was an occasion regarded as one of the "events" of a lifetime, like coming of age or getting married. Farmers far and wide were experiencing the buying fever in its most virulent stage. They usually found that the only way to cure "automobilitis" was to nurse it along and let it take its course. That the motor car was invading the country in 1910 was borne out by these statistics: a total of 26,000 new autos were sold to farmers that year, compared to 4,000 in 1909.

The practical uses being found for the automobile were many. One farmer had this to report in the **Gas Review**:

One of my neighbors recently broke the gasoline engine which had been used for general power purposes on the place. He was running a corn sheller with it at the time. He dragged the damaged engine out of the way, backed his motor car into its place, elevated the rear end, adjusted the belts to the wheels, and the way he turned out shelled corn the remainder of the day was a wonder.

Another farmer, writing to the same trade journal, gave this testimonial:

I make my machine do all the work of a light wagon. Of course I cannot haul loads as heavy as a good team, but I make up the deficiency in the speed and the number of trips possible. I have no horses to feed, to hitch and unhitch on every trifling occasion. I turn the crank and am off. The initial cost was not so much greater than a good team and light wagon and the comfort is double . . . I believe it is the best thing that ever came to the farm.

By 1911 the question of supplanting the farm horse had been discussed for a decade with more or less heat. But while some 500,000 of their numbers had already been displaced by the automobile, the future of these animals remained uncertain. Rather than diminishing in numbers, horses and mules were increasing at the rate of one-third million each year.

International Harvester built this 30-horsepower automobile in 1910 and 1911. After about 1,500 were produced, the company decided to abandon its manufacture. It was decided that trucks were more in keeping with the IHC policy of meeting the nation's transportation needs.

Although use was often limited by road conditions, the automobile was becoming a major social factor in the United States. It was being compared with the telephone, rural mail delivery, and the consolidated school in giving farmers the advantages of urban as well as rural life. It was helping to bring rural and urban people together and to make social communication between them easier.

Many famers believed they were as entitled to the luxury of "joy riding" as were their city cousins. Motorized family outings were becoming quite popular throughout the countryside. One fellow stated that he did not expect the purchase of an auto to be a means of adding to his income; but because he had worked earnestly and continuously for several years, owned a good farm and had money in the bank, it was about time that he should have some fun.

Above: Another favorite of its day was the Chalmers-Detroit. Its promoter, Hugh Chalmers, was a genius in the field of advertising and salesmanship. This 1910 "30" roadster with bucket seat sold for $1,500.

Below: In 1908, Barney Everitt, along with William Metzger and Walter Flanders, formed the E.M.F. Company. Despite remarks by envious competitors, who claimed the letters E.M.F. stood for "Every Mechanical Fault," it was an excellent car in its day. This 1912 "30" roadster sold for $1,100.

Buick is the oldest division of General Motors, having been the cornerstone on which this industrial giant was founded in 1908. This 4-cylinder Model 43 of 1912 was said to "assure the acme of passenger comfort on all roads and at all speeds."

A farmer in Kansas, writing to the **American Thresherman**, summed it up this way;

My experience has been, and I have driven a $2,000 car for two years, that one must expect the pleasure and healthfulness of the sport to make up for part of the expenditure. There is no reason why a farmer of fair means should not be willing to pay the value of twenty or twenty-five steers for five or even ten years of pleasure. I for one am willing to do it even were the motor car of no practical use.

More and more farmers were attending the annual winter automobile shows. Perhaps none of these shows generated more enthusiasm than those of 1913, when the slogan was "All hail the self-starter." With the possible exception of electric-lighting systems, the self-starter was easily the best talking point of the season. There always had been an army of motorists nursing broken arms and legs, plus numerous minor bruises, all from cranking accidents. Others were maimed when run over or sandwiched between radiators and telegraph poles. Then there was the mental distress imposed when the motorist had to "crank up" while standing in water or mud.

The Ford Model T was dominating the automobile market in 1913. Sales for the year exceeded 168,000 vehicles. The runabout was priced at only $525, and this included a long list of equipment.

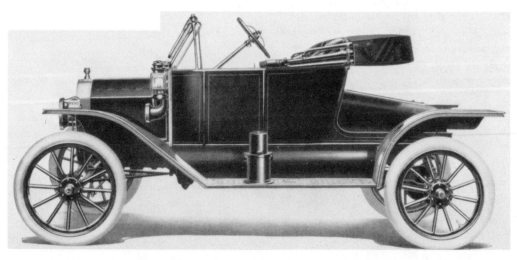

Ransom E. Olds, who was manufacturing the curved-dash Oldsmobile as early as 1901, later began production of a popular automobile which derived its name from his initials — REO. Stating flatly that he had done his very best, Mr. Olds proclaimed the 1912 Reo to be "The car that marks my limit."

An automobile introduced by the Dodge Brothers in the fall of 1914 was the first car in the industry with an all-steel body. John and Horace Dodge had started their successful careers years earlier as manufacturers of engines for Henry Ford. Pictured is a Dodge Brothers touring car in its first production year.

The Oakland, which became a unit of General Motors in 1910, had considerable success over the years. Its slogan was "The Car with a Conscience." This 1914 four-cylinder Model 43 cost $1,785, fully equipped.

More than 1,500,000 automobiles were on the highways of the United States in 1914. About 350,000 of these vehicles were owned by farmers. But horse-drawn carriages continued to be manufactured in great numbers. From July, 1913, to July, 1914, 900,000 carriages were produced, with an estimated value of $54,000,000.

In 1910 nearly all the nation's automobile-owning farmers "put up" their vehicles during the winter months. Motoring at that time of year was beset with difficulties and unpleasantness. The general feeling was "there are only a few days when the roads and weather are good enough for a spin, so what's the use of bothering with the car?"

The vehicle was placed in a shed or barn, where it was jacked up and put on blocks. The tires were removed, wrapped in cloth covered with powdered soapstone, and stored in a dark, dry location. The radiator and gasoline tank were drained, bearings were examined, and various parts were disassembled for inspection and lubrication. Come spring, it took a full day to get the auto in running order again.

Many farmers laid up their automobiles simply to escape the bother of looking after them. Frequently it was because they were afraid they would forget to drain the radiator some night, allowing the engine to freeze. Considering this viewpoint absurd, one dauntless winter motorist remarked: "A person might just as well say that he is afraid to take his horse out in cold weather lest he forget to blanket him when stopping, and thus cause his death through pneumonia."

Motorists were often averse to using any of the non-freezing solutions then available, claiming that they corroded the cooling system. They preferred to allow the engine to idle slowly when a car was not in motion. If the engine was stopped for an hour or two, a blanket or lap robe thrown over the radiator and hood retained enough heat to enable it to start readily. At night when the vehicle was left in an unheated shed or stable, it was necessary to drain the water to prevent freezing.

By 1915 the storage of automobiles during cold weather was no longer a common practice. Side curtain design on most vehicles had been improved so as to open with the doors. In addition, closed cars were making wintertime driving much more comfortable. Considering the inconvenience and discomforts of horse-drawn travel, most car owners had decided their vehicles could be as serviceable in winter as in summer. One farmer reported there were not over a half dozen days each winter when he could not use his car. By employing a storm hood there were no ill effects from exposure, and he was frequently the first to break the trail to town.

In 1917 an estimated 3,500,000 motor vehicles were handling a greater volume of passenger transportation than the combined service of all steam and electric railroads. Highways were no longer a safe place for chickens or pedestrians. With wheat selling at $2 per bushel and pork at $18 a hundred, farmers bought autos as never before. Each purchase not only added another recruit to the list of motoring enthusiasts, but also inspired neighboring farmers to possess a car. Carriage manufacturers began closing their doors. The cost of a new auto was often not appreciably more than for a horse, harness and buggy. And it had four or five times the range of travel.

Above: In addition to farm equipment, J. I. Case for several years also manufactured a line of motor cars. Price for this 1916 "40" four-cylinder touring car was $1,090.

Below: By 1917 the Ford Model T was taking on some semblance of streamlining, although the mechanical units remained basically unchanged. Price for the touring car was reduced to $360, and production reached an all-time high of 750,000 vehicles. One-half of all the cars on American roads were Fords.

When the United States entered World War I, Willys-Overland was the second largest manufacturer of automobiles in the country. This remarkable accomplishment was due largely to the foresight of the firm's president, John N. Willys. The four-cylinder touring car of 1917 was priced at $850, f.o.b. Toledo.

The Dixie Flyer automobile was built for several years by the Kentucky Wagon Manufacturing Company, a Louisville firm that was established about 1880. This circa 1917 roadster featured a "one-man" top and a four-cylinder Herschell-Spillman motor.

The lowest priced car on the market sold for about $395 and the highest for $6,000. The average for all was a little more than $1,600. As to the choice of engines, the field was limitless as far as cylinders were concerned. There were 7 twelves, 25 eights, 63 fours, and 81 sixes for the buyer's consideration, exclusive of the two or three steam cars and the electrics.

Older automobiles, having served their time as passenger vehicles, were often retained on the farm for other purposes. By jacking up the rear end, removing the tires and doing a bit of improvising, power could be had for sawing wood, grinding feed, and numerous other jobs. A mechanic in Kansas City invented a device which could be attached to the front of a Ford "T" in less than 15 minutes, converting it into a stationary power plant. From six to eight horsepower could be produced without boiling the water in the radiator. A company in Illinois offered a steel-wheeled contrivance that fastened directly to the auto chassis, which was said to make "a practical tractor out of a Ford or most any other car."

The livery stable had long provided an essential service for people from many walks of life. A number of horses, carriages, and wagons were always kept on hand and were available for hire. Farmers from outlying areas often relied upon these facilities to furnish shelter and feed for their teams when visiting in town. By 1920 the use of horse-drawn carriages had declined greatly, and livery barns were rapidly giving way to automotive garages. The stench of the occasional manure pile would soon be replaced by the wide-spread pollution of internal combustion.

The old-fashioned livery stable always had been a favorite place for spending leisure time. Being fond of man's noblest friend, the horse, many a boy was attracted there. However, the fraternizing that sometimes took place between boys and the disreputable "characters" who also frequented the stable tended to arouse the consternation of parents. It was thought the public garage would promote higher principles in growing manhood.

Practically every man raised on the farm was familiar with that institution called the blacksmith shop. With the expenditure of strength and elbow grease, many had pumped the bellows or turned the big, cumbersome grindstone.

But now, like the livery stable, the blacksmith shop was undergoing changes. Many of them were becoming repair garages for automobiles, and instead of a string of horses around a hitching rack waiting to be shod, cars were to be found in various stages of repair. And in the material stock where horseshoes, wagon hubs, felloes and spokes once dominated, there were found rubber tires, ball bearings and a confusing array of auto accessories.

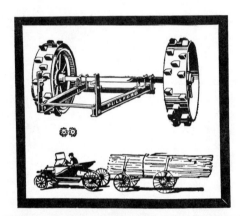

Let Your Ford do the Work

Above: The Apperson brothers, Elmer and Edgar, manufactured a successful automobile for many years. During World War I, they introduced an eight-cylinder car they called "The Eight with Eighty Less Parts." Model of 1920 is shown.

Below: The world's largest manufacturer of wagons, the Studebaker Corporation, for years also held an enviable position in the automobile industry. Its "Special Six" touring car of 1920 featured a 50-horsepower engine. Price was $1,785.

MITCHELL

Five Passenger Sedan

F MADE to individual order, this Sedan could express no greater elegance, feature no more superb furnishings or present a greater degree of substantial dignity.

The luxuriously deep and extra wide seats are upholstered in rich velvety velour. The doors, that can be locked, have receptacles for ladies toilet requisites and gentlemen's smoking sets. Floors have Wilton carpets and heater set flush.

The wide silk curtained windows slide in rubber at the touch of a finger. Dome and reading lights are of chased silver.

The same Refracting Windshield as on Coupe, spells safety for night driving while it emphasizes the smart style lines that give this car such character and distinction.

This Sedan is the ideal all season's car for family use and for the man of affairs. Women, particularly, admire Mitchell Sedans as well as Coupes, for their ease of control and unusual roominess.

Young "Abe" Lincoln was postmaster at New Salem, Illinois, in 1834 when Mitchell was founded.

The Mitchell automobile was manufactured by the Mitchell & Lewis Company, a firm long famous for its wagons. This six-cylinder, 40-horsepower sedan of 1921 was one of the last Mitchells built.

The Maxwell, dating from 1904, was a popular automobile for more than two decades. In later years its slogan was "The Good Maxwell." This four-cylinder roadster of 1920 appealed to many because of its sporty lines.

Chevrolet production in the early 1920's was insignificant when compared with Ford, but this would soon be changing. With "fixed seats for three and a comfortable folding seat for a fourth passenger," the Chevrolet coupe of 1922 was obviously quite commodious.

149

Above: The Star captured some of the low-priced automobile market for a period in the 1920's. Its creator, William C. Durant, was a former manufacturer of carriages. In 1922 this Star four-door sedan could be purchased for the rather nominal sum of $645.

Below: The Hudson Motor Car Company once manufactured a type of low-priced closed car which it named the "coach." This Essex coach of 1923, built by Hudson, offered "closed-car comfort at open-car price."

Although steam-powered automobiles cut a considerable swath during the early years of the 20th century, they gradually lost the battle for supremacy to those using gasoline. Motorists began to wonder why they sould bother with pilot lights and steam gauges and why they should spend 15 or 20 minutes "firing up" before a vehicle could be driven. Then, too, it was necessary to know where all the watering troughs were located if traveling very far from home. The Stanley, shown shortly before its demise in 1925, was the most successful steam car ever built.

Production of the Ford Model T ended in June, 1927, after more than 15,000,000 vehicles had been built. Wire wheels were standard on the 1927 sedan and also on all other models.

The Nash, a descendant of the once-famous Rambler, was first manufactured in 1918. The 1930 sedan featured a "new high-compression motor capable of speeds of from 65 to 70 miles an hour."

A letter, written by a farmer in 1916 and published in the **Gas Review**, typifies the sentiments that prevailed during the transitional years of American transportation:

Dear Billy:

Yes, we took your advice and purchased a machine last month. It is a dandy, runs with a nice soft musical hum and takes us down to the village and back in one-half hour, just two hours less than old Frank and Dorothy needed for the same trip. You remember how prejudiced I was against motor cars? They used to drive up behind me with those snarly horns which sort of swear at you and seem to say "Get out of the road, you lobster." I hated those horns. We have one of the same kind on our car.

I never used to envy the motorists that ran down past our farm. Their women folks never looked comfortable in the back seat, the wind blowing their mouths full of hair as they gazed down the road with their eyes blinking and their jaws set. Ma says she likes it; however, she misses the horse hairs on her Sunday suit when we go to the meeting house and it doesn't take her all day to dust off her skirt. Yes, Ma likes the car and expects to learn to drive it herself.

You know, Billy, I have been using our old plow team for family driving long enough. Many a time on Sunday I have hitched them to the surrey after a hard week of work and have driven them down to the village church feeling so ashamed of myself I knew I ought to have walked or stayed at home. Frank and Dorothy had earned a big holiday, but some way we wanted to get there and the horses had to stand for it. Well, the crops were pretty good this year, wheat ran ten bushels to the acre more than I expected and Ma suggested that we buy a nice little driving horse and a new buggy. We figured out the whole deal, horse two hundred dollars, buggy one hundred dollars, and then began thinking about a new harness. I had never figured on owning an automobile as long as I lived, they seemed like a luxury and you know that Ma and I have dreaded luxuries like the devil and

poison. After losing two nights sleep thinking about it, we decided to buy a little car. It has only eaten about three dollars worth of gasoline since I began keeping track last week and it has been doing a lot of business without any currying or hay pitching to be bothering us each morning and night.

They held a farmer's meeting over at John Adam's three days after I bought the car. Ma said, "Stay at home, you don't know enough about running the thing." I said "No, I will squeeze one pedal with my foot and if it don't work I'll try another." The blamed machine works right or it stalls dead in the road so you either steer and keep going or you get fussed and stop. Perfectly safe either way you do it.

Last week the binder broke down while we were busy in the field, our hired man was a bit careless, but I cranked up and drove down to the blacksmith shop and in less than two hours we were using that binder again. It was a fine day, too, and we can't afford losing much time when good weather happens to come around for a brief spell.

Why, yesterday I laid back the top and folded it according to directions and we had a nice big seat almost like a delivery wagon. I just loaded up those three chicken crates with the old hens that we had been keeping two weeks too long for want of time to haul them to town, and in half an hour after catching them with the hen crook they were down to the express office and by next week there will be a nice little check from the commission man down at the city.

They way automobiles are an expense with gas going up and tires bursting down but the little bus is helping me to earn more money on the farm, at least it's keeping me on the job earning as much as I used to, and I was getting ready to leave the place. If I'd left I never would have earned a red cent any other place; a farmer can't get onto the ropes in the city after he gets my age.

I always used to worry about a gasoline engine. I didn't know why they went when they went and why they stopped when they stopped. I have been reading the instruction book every day during the last three weeks and I find that the makers knew enough to put it together right so I haven't altered it any. They advise oiling, so I oil, and every grease cup gets stuffed full whenever we get ready to go. When our little car backs out of the old red carriage shed, she is greased up to slip along nicely, and she slips. We haven't had any trouble yet and she runs smoother every day.

Billy, my boy, your old uncle don't blame you for going to the city; he used to want to go himself, that is, up to three weeks ago. We used to feel isolated sometimes, Ma and I, especially in the evening when the church bells rang faintly across the marsh over beyond Aker's woods and the old horses were too fagged to make it a decent act to hitch them up. And then I was tired, too, on those nights and it was no small job to take all of that trouble even for the meeting, especially after finishing a full day in the field and looking forward to the next day which seemed bigger and longer. Now we crank up the little car and we soon get there and when we are ready to come home, we come, and the whole event isn't any more trouble than visiting your neighbor a couple blocks down the avenue in the city.

I used to fairly cuss these city fellows who came rambling across the country in their cars, looking so fresh and sort of aristocratic, but I have decided that they are just like Ma and me, out for an airing and trying to forget some of their pesky troubles. I forgive the blond-haired fellow that killed my pig last year and will forget about the old speckled rooster which was all messed up one Sunday afternoon down the road along the pear orchard.

Billy, the car has made a new man out of me, it seems like something alive and it minds better than any old horse we ever had on the place. We are getting out and seeing the world now. It is a much better place than we thought it was; just our township has a lot of good things that I never dreamed of and we have visited friends that I had nearly forgotten. Come and see us during your vacation this year. The old horses will not bore you, and all ready for the twist of the crank the little car will be waiting at the station ready to slide along with that friendly hum which will carry us out to Ma and the old place again.

As ever,

Uncle Abe.

This 1930 Buick Four-Passenger Special Coupe was a fashionable automobile in its day. In its spacious rear deck was a large rumble seat, finished in genuine leather, designed to carry two extra passengers "in perfect comfort."

SELECTED BIBLIOGRAPHY

BENTLEY, John. 1952. The oldtime automobile. Fawcett Publications, Inc., Greenwich, Connecticut.

BRUMFIELD, Kirby. 1974. The wheat album. Superior Publishing Company, Seattle, Washington.

BRUMFIELD, Kirby. 1968. This was wheat farming. Superior Publishing Company, Seattle, Washington.

CLYMER, Floyd. 1955. Henry's wonderful Model T. McGraw-Hill Book Company, Inc., New York, New York.

CLYMER, Floyd. 1953. Those wonderful old automobiles. Bonanza Books, Crown Publishers, Inc., New York, New York.

CLYMER, Floyd. 1950. Treasury of early American automobiles. McGraw-Hill Book Company, New York, New York.

GLASSCOCK, C. B. 1937. The gasoline age. Bobbs Merrill Company, New York, New York.

GRAY, R. B. 1975. The agricultural tractor: 1855-1950. American Society of Agricultural Engineers, St. Joseph, Michigan.

HOLBROOK, Stewart H. 1955. Machines of plenty. Macmillan Company, New York, New York.

JOHNSON, Paul C. 1978. Farm power in the making of America. Wallace-Homestead Book Company, Des Moines, Iowa.

KAROLEVITZ, Robert F. 1966. This was trucking. Superior Publishing Company, Seattle, Washington.

MEYER, Henry W. 1965. Memories of the buggy days. Brinkler, Inc., Cincinnati, Ohio.

RITTENHOUSE, Jack D. 1951. American horse-drawn vehicles. Floyd Clymer, Los Angeles, California.

SCHLEBECKER, John T. 1975. Whereby we thrive. Iowa State University Press, Ames, Iowa.

THE YEARBOOK OF AGRICULTURE. 1960. Power to produce. U.S. Department of Agriculture, Washington, D.C.

WIK, Reynold M. 1953. Steam power on the America farm. University of Pennsylvania Press, Philadelphia, Pennsylvania.

PERIODICALS CITED

AMERICAN THRESHERMAN. February 1904-November 1913. Madison, Wisconsin.

GAS REVIEW. January 1909-December 1917. Madison, Wisconsin.

THRESHERMEN'S REVIEW. May 1900-September 1911. St. Joseph, Michigan.

THRESHERMEN'S REVIEW AND POWER FARMING. June 1914-May 1915. St. Joseph, Michigan.

TRACTOR AND GAS ENGINE REVIEW. January 1918-April 1925. Madison, Wisconsin.

NEWSPAPERS CITED

ASHLAND PRESS. January 1890-December 1919. Ashland, Ohio.

ACKNOWLEDGEMENTS

Donald Anderson, Grafton, Ohio.

John and Lee Blosser, Orrville, Ohio.

Dale Fasnacht, Massillon, Ohio.

Donald M. Irvin, Creston, Ohio.

Herman Kasten, Springfield, Ohio.

Forest Smith, Ashland, Ohio.